A Practical Guide to
Designing Expert Systems

A Practical Guide

to

Designing Expert Systems

Sholom M. Weiss
and
Casimir A. Kulikowski

Rowman & Allanheld
PUBLISHERS

ROWMAN & ALLANHELD

Published in the United States of America in 1984
by Rowman & Allanheld, Publishers
(A division of Littlefield, Adams & Company)
81 Adams Drive, Totowa, New Jersey 07512

Library of Congress Cataloging in Publication Data

A pratical guide to designing expert systems.

Bibliography: p.
Includes index.
1.   Expert systems (Computer science) 2.   System
design. I. Kulikowski, Casimir A. II. Title.
QA76.9.E96W45 1984   003   83–27040
ISBN 0–86598–108–6

84 85 86 / 10 9 8 7 6 5 4 3 2 1
Printed in the United States of America

Contents

Tables and Figures

Preface

More than ten years ago we implemented our first large scale expert system, CASNET, demonstrating the feasibility of a computer-based model of expert reasoning for medical diagnosis. When we began that project most artificial intelligence (AI) research was concerned with generalized strategies of problem solving, taking little advantage of domain-specific knowledge. There were few exceptions, most notably the DENDRAL project. Today, the change in perspective is drastic indeed. Tackling real-life problems with realistic models of reasoning is now the norm rather than the exception. Research on knowledge-based systems like CASNET/Glaucoma, MYCIN, PROSPECTOR, INTERNIST-I, and R1, made this change possible.

Formalizing the experience of human experts and making it scientifically testable is one of the major goals of expert systems. We have tried to do the same in this book, gathering our own experiences and formalizing them after a fashion. Aware of other references in the field, we have concentrated on our own approaches and results. These are best described as dealing with a well defined class of expert systems—those that solve classification problems using a rule-based approach. A classification problem is one where we must place a subject, object, or phenomenon (signal) into one of several prespecified classes. Diagnostic decision-making in medicine is one of the clearest examples of a classification problem. Other examples are the interpretation of geological signals in mineral prospecting and oil exploration and the selection of advice in repairing faulty equipment. Rule-based approaches are those that use a large store of rules for their reasoning. A major point of this book is to show that representing expert knowledge need not be very complex; strategies of reasoning can be designed to work without excessively long or arcane chains of logic. A reliable expert, a knowledge engineer who will debrief the expert, and a good sample of test cases of difficult problems are the elements required to design and then to challenge an evolving expert system.

In approaching the design of an expert system, we have found that good programming skills are helpful, but usually secondary to knowledge of how decisions are made in a problem domain. Many of our graduate students have succeeded in developing prototypical expert systems because they were willing to commit themselves to learn about subjects such as medicine or geology that were far afield from computer science. Such commitments have to be sufficiently strong to gain an expert's trust that his knowledge will be expressed reliably in the computer model built by the knowledge engineer. We have written this book for those who would like to help build an expert system, but do not necessarily have experience in AI programming.

There are many people who have helped us over the years, and have contributed either directly or indirectly to the results contained in this book. Peter Politakis, John Kastner and Chidanand Apte have contributed substantially to our efforts in system design. Kevin Kern has programmed the software supporting many of our systems. Rich Keller and Bob Goldberg have contributed to review materials in Chapter 3. Dick Duda made numerous valuable comments on an early version of the manuscript. Angela DiCorrado has been most helpful in typing various parts of the manuscript. We thank them, our families, and the National Institutes of Health, which have supported our research for many years.

1

Expert Problem-Solving and Consultation

1.1 What Is an Expert System?

An *expert system* is one that:

- handles real-world, complex problems requiring an expert's interpretation
- solves these problems using a computer model of expert human reasoning, reaching the same conclusions that the human expert would reach if faced with a comparable problem.

Experts are people who are very good at solving specific types of problems. Their skill usually comes from extensive experience, and detailed specialized knowledge of the problems they handle. Examples include consulting physicians, interpreters of oil well data, and engineering experts who carry out diagnosis and repair of high technology equipment, such as computers.

A computer-based *expert system* seeks to capture enough of the human specialist's knowledge so that it too will solve problems expertly. Over the past ten years various research groups in *artificial intelligence (AI)* have built highly specialized systems containing the expertise needed to solve problems of medical diagnosis and treatment, chemical structure analysis, geological exploration, computer configuration selection, and computer fault diagnosis, among others. Table 1.1 lists a representative group of expert systems as of late 1983. Although these systems are very different and specific to each application, it has gradually been discovered that computer-based techniques for representing knowledge and reasoning with expertise can be quite general.

Practical applications for expert systems abound. Whenever human experts are in great demand and short supply, a computer-based consultant can help amplify and disseminate the needed expertise. An expert system can capture the practical experiential knowledge that is hard to pin down and that is

Table 1.1 Expert systems having verified performance or undergoing verification—1983

Expert System	Application	Developed by
Serum Protein Diagnostic Program [a][b]	serum protein electrophoresis	Helena Labs/Rutgers
PUFF [a][b]	pulmonary function tests	Stanford/Pacific Medical Center
CADUCEUS/ INTERNIST-I [b]	interal medicine diagnosis	U. of Pittsburgh
MYCIN [b]	antimicrobial therapy	Stanford
CASNET [b]	glaucoma diagnosis/therapy	Rutgers
R1 [a]	VAX computer configuration	DEC/CMU
PROSPECTOR [b]	geological exploration	SRI
DENDRAL [a]	mass spectroscopy	Stanford
MACSYMA [a]	symbolic integration	MIT
DART/DASD [b]	computer fault diagnosis	IBM/Stanford
ELAS[b][c]	well-log interpretation	Amoco/Rutgers
Dipmeter Advisor[b][c]	dipmeter well-log interpretation	Schlumberger
ACE[b][c]	telephone cable maintenance	Bell Labs

[a] extensively used
[b] classification-type system
[c] currently undergoing field testing

rarely, if ever, found in textbooks: the alternative methods by which an expert solves specific problems. Let's examine three sample problems where an expert system may be advantageous.

In *medicine*, seeking a second opinion for complex or hard-to-treat problems is becoming increasingly important as specialization grows. The proliferation of new instruments and laboratory tests, makes keeping up to date with the technical and clinical details of their interpretation a taxing job for the busy practitioner. Summarizing the expertise of the best specialists within computer models can provide a useful touchstone by which the latest results of medical research are put at the fingertips of the clinician.

In *oil exploration*, there is a shortage of expert well-log analysts, and as result they tend to move from company to company as they are offered progressively better financial inducements. From an oil company's viewpoint, the possibility of capturing some of the expertise of the best analysts in a computer model becomes a very attractive alternative for training new specialists and retaining existing proven expertise. And the very process of building an expert system makes the human expert record his reasoning in an explicit and formal representation that clarifies and opens the way for improvements in the decision-making.

In sophisticated *equipment repair*, such as that needed for repairing computers, there is likewise a shortage of trained experts, and as technology becomes more complex, this problem will get progressively worse. The dynamic components of expertise are very hard to capture in conventional manuals, and even if they can be captured in a manual, one must still decide which section of the manual is appropriate for a particular problem.

We believe that there are already known techniques for building practical and effective expert systems. Although currently there are not many expert systems in everyday use, we have seen the power of these techniques when, for example, we were able to take an interpretive program for one of the most widely used and interpreted laboratory tests in medicine—serum protein electrophoresis —from inception to production in less than a year. This is a highly circumscribed and relatively small problem. Researchers, however, are using similar methods in a variety of complex applications: oil and mineral exploration, equipment repair, signal interpretation in military situations, and in many medical diagnostic problems.

An understanding of how the human mind actually works in solving expert problems is not necessary to successfully produce the expert systems that will augment human capabilities and productivity. It is sufficient to be able to *debrief* an expert of his or her knowledge, and structure this knowledge in a uniform computer representation that will permit the application of consistent methods of processing on the computer. This does not mean that any problem requiring expertise is amenable to computer solution using the current techniques of expert systems. Rather, a very broad class of problems can be effectively solved by these methods.

This book is intended as a practical guide for people who would like to learn about the first generation of expert systems techniques that produce working programs. We deliberately restrict our coverage to techniques that are reasonably simple, yet effective. The class of problems that we deal with is also well defined. We consider the very broad category of *classification* problems. Typical classification problems are diagnosis, interpretation, or giving advice.

Examples of expert system design tools that can be used to encode a classification-type model include *EXPERT, EMYCIN*, and *PROSPECTOR*. *EXPERT* has been developed based on ten years of experience with medical consultation programs, and more recently oil exploration, computer fault diagnosis, and a variety of other applications. *EMYCIN* is a generalization of the MYCIN system; MYCIN gives advice on the treatment of infectious bacterial diseases. The *PROSPECTOR* system, and its associated knowledge acquisition system *KAS*, has been developed using experience gained with geological prospecting models. Although there are a variety of other formalisms that could be used for encoding expert knowledge, we emphasize these three as typical representatives of classification-type *rule-based* systems, which encode the knowledge critical for decision-making in the form of highly modular reasoning rules.

We will draw most extensively from our own experiences. As a result, what follows is a somewhat biased narrative. We concentrate on the *software engineering* aspects of designing and implementing an expert system. More extensive discussions of the philosophy, comparative approaches, and cognitive issues are available in other existing and forthcoming publications. Such topics are necessarily more tentative than the engineering ones because of our limited understanding of human cognitive processes. We recognize that these topics are important, but we do not feel that their consideration should hold back the more empirical engineering efforts to build expert systems that work, even if only in highly specialized areas.

1.2   Two Examples of Expert Systems

In this section we will briefly touch on two projects with which we have personal experience, showing how ideas that two or three years ago might have seemed difficult if not impractical, are now realized as practical expert systems.

1.2.1   EXPERT INTERPRETATION OF LABORATORY INSTRUMENT DATA

With the proliferation of tests in a modern clinical laboratory, much time is spent by the physician in differentiating normal from abnormal results, cutting into the time available for the detailed analysis of the abnormal cases. Delays in interpreting results can occur, with possible harmful effects on patients. Expert clinical pathologists often develop criteria for distinguishing normal from abnormal test results that are explicit and codifiable within an expert reasoning model.

Working with Dr. Robert Galen, a leading clinical pathologist, we developed a small expert system for the analysis of serum protein electrophoresis, an important laboratory blood test. It also presented an opportunity for increasing productivity in the clinical laboratory, where a large number of many different tests are carried out daily, requiring review, interpretation and the generation of reports by the pathologists in charge. These reports are then sent to the patient's primary or referring physician, who will integrate the results with the other facts available about the patient's condition.

The serum protein electrophoresis (SPE) laboratory test has been shown in some surveys to be the most frequently interpreted test in the clinical laboratory. The instruments on the market at the time we began this project had digital output for the results of the test, but no interpretive reporting. While the interpretive reporting of laboratory tests was part of a growing trend in hospital or clinic systems, none had as yet been integrated into an instrument. We felt that in addition to the practicality of this application, an instrument-derived interpretation would probably be more acceptable as a natural extension of the instrument rather than a free-standing consultant program. Many recent medical instruments are microprocessor controlled, and we knew it would be relatively easy to add an interpretive component to enhance the performance of such a device, at relatively little extra cost to the manufacturer.

In selecting the interpretation of serum protein electrophoresis (SPE), we deliberately chose a problem that was circumscribed and provided a routine interpretation task for pathologists, yet had enough nuances that complex cases would require expert advice.

Using a generalized expert systems approach, in six months, we were able to develop a program that produced an interpretive report directly from the output of this instrument. The interpretive program was eventually placed in a ROM (read-only memory) and distributed as part of the instrument to several dozen medical laboratories. The technical approach to designing this expert system, with an example of the instrument's interpretation, is discussed in Chapter 5.

1.2.2 EXPERT INTERPRETATION OF OIL EXPLORATION DATA

In the early days of oil exploration, the primary way of knowing whether oil was present was to drill a hole and see if oil gushed out. The chances of actually hitting the right spot in a reservoir that would produce a gusher became progressively smaller as the well-established geological locations for oil were intensively developed. By the 1920s the need to predict the presence of hydrocarbon deposits (oil, gas, and their various mixtures) that were not just close to the surface was sufficiently great that a number of methods were developed. Among the most conclusive is well-logging, which is a method of sending sensors down a hole and recording the signals transmitted by the sensors.

The well-log analyst is the person who interprets the well-log data in the light of the location's known geological features and then estimates the likely presence and amount of hydrocarbon deposits. Expert well-log analysts are highly prized in the oil industry. They often move from company to company, and their expertise may be, at first glance, difficult to put on paper, consisting of their experiences in interpreting well-logs from many oil fields around the world. However, just as in medicine a physician abstracts rules of reasoning or criteria for analyzing patient data, so well-log analysts have certain typical ways of proceeding through their analysis of well-log data.

Because of these similar characteristics, it is possible to consider a well-log analysis problem as one that fits into the classification framework and then proceed to apply expert systems techniques to encode the human expert's knowledge.

Our experience in this field comes from a collaborative project with the Amoco Production Company, which in recent years has been one of the largest driller of wells in the continental United States. In developing an expert system we wanted to use existing analytical software components while building an advice-giving and problem-solving system that would facilitate and guide their proper use. In this oil exploration problem we are not only trying to extract the expert's reasoning about the interpretation of data, but also the many alternative paths that can be taken in choosing how to use the different analytical techniques that are available on the system. For the advice to be truly expert, the expert system must be able to monitor the path taken by a user, and it must offer interpretations and suggested courses of action. This requires that the logic of the system be designed to interpret evidence in a real-time situation, so that dynamically changing choices on the part of the user are always reflected in the conclusions of the model.

Working with Jay Patchett, a leading well-log analyst, we have developed a system called ELAS (Expert Log Analysis System). This system will be described in Chapter 5. The example of ELAS shows how an expert system can exploit existing software by coordinating and advising on its use. Given the expense of developing applications software and the desirability of using it effectively, this promises to be an important class of applications for problems that can be cast in the classification mold.

1.3 Structure of Knowledge and Reasoning in Expert Systems

Several characteristics typical of diagnostic or interpretive expert systems are exemplified by the two applications described in the previous section. In particular, we can draw some generalizations about the structure of knowledge and the forms of reasoning that they share.

- *The conclusions* (or advice) must be generated by the system from a finite set of discrete and prespecifiable elements. For example, all qualitatively different types of dysfunctions, or modes of breakdown, of a piece of equipment can be listed by the experienced diagnostician.
- *The evidence* (observed facts or data) about the problem must be reliably obtained by the user of the system or by the system itself.
- The *initial assumptions* must limit the problem to a highly specialized area, thereby directing the reasoning.
- A *knowledge base*, or structure of relations and reasoning rules, must be available to link the evidence about the problem to the conclusions.

These may be logical or probabilistic rule-based associations.

- A *reasoning control strategy* must be designed to guide the reasoning of the system and make its output behavior correspond to acceptable sequences of responses for its human users. For instance, it is usual for an expert to interrupt the solution of a problem and ask for additional data, if necessary. An expert system must in some way know when to do this, and just like a human must give its reasons if they are solicited. If certain data are unavailable, the expert system must be able to try alternative methods that rely on different available data.

Figure 1.1 illustrates the major components of an expert system and points out how the problem-solving can be viewed as a flow of reasoning that goes from evidence to conclusions. The goal is to come up with the appropriate advice or conclusion for a particular case. Much of the skill of human experts lies in their ability to narrow down the scope of possible solutions with every additional piece of evidence that is received. In the same way, an expert system with a well-designed reasoning model will make use of many different patterns of evidence, combining them into intermediate hypotheses about partial conclusions from which the final results will be eventually inferred. The rate at which the field of plausible hypotheses can be narrowed is often quite fast. These diagnostic or interpretive expert reasoning models differ substantially from those for game-playing or design problems where the choice of hypotheses is extremely large, and often it is quite difficult to narrow the hypotheses to a small number. Details of the structure of knowledge and the types of reasoning and control methods will be discussed in Chapter 2.

1.4 Motivations for Building an Expert System

Why build an expert system? We have mentioned the most obvious reasons: dissemination of rare and costly expertise, and the more effective and efficient use of the human expert. Other reasons go beyond these productivity considerations.

From a scientific point of view, the most important reason is the formalization and clarification of knowledge that results from having the human expert make his reasoning explicit. While specialized texts often capture the details of how phenomena are explained and understood, rarely if ever do we find details of how to apply this knowledge in an expert fashion. Human experts, such as physicians, well-log analysts, and engineering specialists, have many state-of-the-art methods based on their experience with many difficult specific problems. They rarely abstract their reasoning methods and systematically describe how, when, and where they are to be applied. This usually forms part of the informally transmitted lore of the expert, which is picked up by apprentices watching their masters at work. In building an expert system, we can encourage the human expert to put down his expertise in a form that

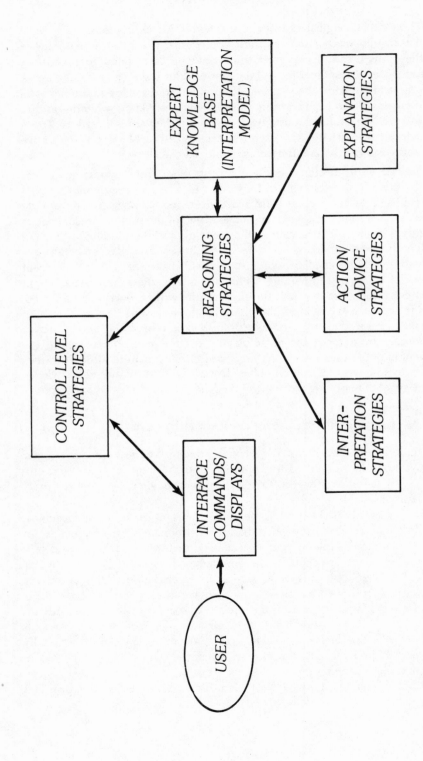

Figure 1.1: Major components of an expert system showing the principal paths of reasoning

will be reproducible and testable by others beyond the confines of his immediate circle.

Another reason for building expert systems is the possibility of combining the expertise from many human experts into a shared knowledge base that can be then studied for consistency and reliability of its advice. From these studies, a synthesis of the knowledge of the many experts could emerge in ways that are new and different. For this to happen, we must develop new methods for comparison of expert reasoning. Above all, we must have ways of exercising and testing the knowledge base against real cases.

In the following sections we consider in more detail the motivations for building expert systems and some of the benefits that they entail.

1.4.1 DISSEMINATING RARE AND COSTLY EXPERTISE

On the face of it, no one would argue with this motive. Saving lives, improving the quality of life, preventing malfunctions, diagnosing them and suggesting remedies are all valuable goals of expert systems. Yet, we must ask about the cost at which this is to be done, and whether there are alternative ways of distributing the expertise of knowledgeable humans. Further, we must ask whether an expert system is likely to be used correctly, and if the potential for misuse might in some way be worse than its complete absence.

First, the issue of cost. Until the 1980s most expert systems were prototypes developed as demonstration projects that ran on large and expensive machines, subsidized by government grants. With the advent of microprocessors and personal workstations, this picture is changing. Expert systems that can run on dedicated microprocessors are now being developed, and the low cost of memory and storage will make it possible to store even some of the most ambitious knowledge bases on the new systems. With systems already on the market, it is possible to make available specialized expert systems, or even general purpose expert systems tools for building and testing them. With machines that are relatively inexpensive, the likelihood that expert systems will be widely used increases greatly.

Among the main reasons for wanting to disseminate expert human knowledge more widely are the twin bottlenecks of the current state of affairs: human experts are in short supply, and when available, have little time at their disposal. While they may be very proficient at their work, the possibility of distraction by many different conflicting demands makes them more vulnerable to errors than a computer-based system. The expert's advantage still lies in the flexibility of human response and the power of human sensory pattern recognition. While computer-based visual recognition and sound recognition are still far from matching that of humans, there is an advantage in developing systems that capture the systematic component of expert problem-solving, but continue to use the perceptual input of people wherever it is superior. Thus, we don't expect expert systems to completely replace humans in most applications, although they might in some. Rather, expert systems may serve as interactive *intelligent problem-solving and advisory* systems that augment the capabilities of the user.

1.4.2 FORMALIZING EXPERT KNOWLEDGE

The advancement of scientific knowledge involves the clarification of how nature works, and how man works with nature. Human experts often work in areas where our scientific knowledge lags behind the practical or empirical knowledge of how to make nature bend to the wishes of man. Experts are practitioners, such as physicians, who work under the imperative of treating a suffering patient in the best way possible, without waiting for a detailed scientific understanding of the disease. In doing so, the physician can later suggest experiments that will support or reject various new ideas about the disease process that he may develop in the course of treating many patients with the same problem. In formalizing the expert knowledge of how an expert human solves difficult problems with today's best knowledge, we are laying out explicitly how future alternatives can be sought. As long as the expert states his reasoning only informally and imprecisely, it is impossible to pin down the alternatives; but as soon as there are formalized statements that enable a computer to reproduce the outcome of human reasoning, we can proceed to experiment and see under what circumstances these statements are applicable. Most textbooks do not capture the expertise of decision-making; they describe associational knowledge. Computer programs can simulate and test decision-making. Where a text can state the facts that enter into decision-making, a computer procedure can allow us to try out the myriad ways in which the facts can be combined to produce the outcomes of expert human reasoning. The expert system thus becomes an empirical tool for experimenting with the representation and uses of knowledge. As such it can make an invaluable contribution to the advancement of practical knowledge.

1.4.3 INTEGRATING DIVERSE SOURCES OF KNOWLEDGE

In any given area of study there will be many different experts, often with diverse opinions on how to solve their problems. Expert systems may be able to help in comparing and judging the alternatives, because they require experts to use a consistent formalization of knowledge in capturing the reasoning rules.

Expert systems have already been used as vehicles for summarizing and synthesizing expertise from various sources in a specialty. Some of the earliest medical expert systems, like CASNET and MYCIN, used knowledge from several experts in building their reasoning models. In building the knowledge base for CASNET we had a national network of investigators to suggest rules for the consultation system. Different experts were responsible for a subspecialty's set of reasoning rules in the knowledge base. As with a human expert, the benchmark of any expert system is good performance, and an expert's knowledge that does not work right with real cases of complex problems must be reconsidered for inclusion in a shared knowledge base. As in the real world, we may find disagreement in our experts' views. If these views result in substantial differences in conclusions, it becomes important to carefully point out in the expert system the authorship of each, and indicate that there is disagreement among human experts in this domain. This does not detract from the performance of the program; rather it enhances the quality of the expert system by offering alternatives.

While the reasoning rules are likely to show some variation for different experts, one of the hopes of researchers in artificial intelligence has been that these differences might be reconcilable in terms of a *deeper model* of the underlying processes which may describe the domain. We, for instance, developed a causal model for characterizing disease processes and used it to carry out reasoning in CASNET. Unfortunately, much human knowledge is of an empirical nature. For example, we often know that a drug controls some disease long before we understand the mechanisms by which it achieves its effect. An expert system, like a human expert, cannot afford to ignore practical recipes that will solve a problem, just because our understanding is imperfect or incomplete.

1.5 The Process of Building Expert Systems

1.5.1 KNOWLEDGE ACQUSITION

The general motivations for building expert systems discussed in the previous section apply to any system of this type, but when we build a specific model of reasoning, the motivations have to be much more precise. We build an expert model to solve a specific class of problems and to give advice on their solution. In this section, we consider briefly some of the issues involved in acquiring expert knowledge to build an expert system.

A human expert has many different types of information to provide the builder of an expert reasoning model, including:

- personal experience of past problems solved
- personal expertise or methods for solving the problems
- personal knowledge about the reasons for choosing the methods used.

Relating individual problem experience to abstracted rules of expertise is rarely an easy task for human experts. Often they will be hard pressed to even describe their expertise in a systematic manner, let alone in a rationally structured form. And to get them to provide a close coupling with generally accepted formal scientific knowledge is often quite difficult.

That *expert systems* have overcome, in a small number of instances, the above barriers is already an accomplishment and one of the reasons for the current widespread interest in them. A key ingredient for the success of those expert systems in everyday use is that expert knowledge has generally been represented as a modular collection of *rules* with relatively well-established and agreed-upon conclusions, or *units of advice*. If we consider more detailed descriptions of human expert reasoning, as might be necessary if we were developing a psychological theory of expert reasoning, we might well have to worry about the complexities of tracing vast numbers of plausible links. With the relatively shallow expert systems approach, we ask our expert model designer to group knowledge of the right size and degree of specificity to solve the problem, without being unnecessarily abstract or overly detailed.

An important question in knowledge acquisition is, who will be building the knowledge base? A *knowledge engineer*, familiar with expert systems methods, is needed with current systems. The knowledge engineer must interview an expert, abstract the main characteristics of the problem, and then proceed to build a prototype system. There are advantages in using one of the existing general expert system tools to provide a framework for the knowledge engineer to experiment and refine the expert system. We may hope that the expert will structure this knowledge in such a way that a relatively untutored programmer can enter it into an expert systems framework. Up to now, our experience shows that this does not represent a realistic approach. Most problems requiring expert advice have sufficient nuances and complexities that it is impossible to expect an inexperienced person to take a rough description from the expert and code it directly.

In summary, the knowledge acquisition process is one of the most difficult phases of expert system building. If one recognizes that the application is a diagnostic or interpretive problem, one of the generalized expert system tools may prove beneficial.

1.5.2   CHOICE OF COMPUTER REPRESENTATION

The choice of computer representation must have two attributes:

- power to express the expert knowledge
- simplicity to describe, update and explain the knowledge in the model.

The *power* of a representation can be measured in several ways. The most important concern is that the relevant expert knowledge needed to solve a problem be relatively easily representable. There is often a trade-off between the ease of representing knowledge in a computer and the richness in structure of possible semantic relations that may be described in an expert systems framework. For instance, if we wish to reason explicitly about the cause-and-effect relationship of certain events, we will require additional structure to be described by the knowledge engineer. While such a system may be more powerful, a reasoning model that uses causal relationships will generally be more difficult to encode.

An important question in choosing a representation should be the ease with which knowledge can be changed and updated. Experts are notoriously fickle in their descriptions, particularly in the early stages of describing a model, and flexibility is essential if we are to keep up with them. An expert system representation that is unwieldy or requires complex structures not easily remembered by the expert or the knowledge engineer will certainly not facilitate the development of a proficient reasoning system. In fact, many early artificial intelligence systems, which used complex descriptive frameworks, are singularly poorly suited for expert systems development.

1.5.3 STAGES OF EXPERT SYSTEM BUILDING

We must recognize that a key to success in building an expert system is starting small and building incrementally to a significant testable system. Empirical validation must be carried out as the various stages of refinement progress.

Various stages in the development of an expert system can be abstracted from our experiences. These represent rough guidelines and should not be taken as the necessary course of development for all expert systems.

A. *Initial Knowledge Base Design Stage*:   This comprises three principal substages:

- Problem Definition: the specification of goals, constraints, resources, participants and their roles.
- Conceptualization: the detailed description of the problem how to break it down into subproblems; what are the elements of each, in terms of hypotheses, data, and intermediate reasoning concepts; how these conceptualizations affect possible implementation.
- Computer Representation of the Problem: a specific choice of representation for the elements identified during the conceptualization phase; the first phase that requires computer implementation. Questions of information flow and articulation of the concepts and data will be raised more completely at this stage.

B. *Prototype Development and Testing Stage*: Once the representation has been chosen, we can begin to implement a prototype subset of the knowledge needed for the whole system. The choice of subset is crucial; it must include a representative sample of the knowledge that is typical of the overall model, yet it must involve subtasks and reasoning that are sufficiently simple to test. Once the prototype produces acceptable reasoning, it can be expanded to include more detailed variants of the problems it must interpret. Then it will be tested against more complex cases that will be used as a standard test set for subsequent refinement of the knowledge base. Many adjustments of the primitive elements and their relationships are bound to come about as the result of this testing.

C. *Refinement and Generalization of the Knowledge Base*: This stage can take a considerable amount of time if we expect to reach very expert levels of performance. However, it is sometimes possible to get quite proficient performance in circumscribed specialties after only a few months of effort.

The process of expert-model building has been outlined in this section, and further details tied to specific examples will be given in the remainder of the book.

1.6   Outline of the Book

This book is intended as a combination "how to" and "what is" guide to expert systems, with emphasis on the "how to" part. This is an engineer's book and not a research review. We have therefore concentrated most of our efforts to describing concrete systems, frameworks for representation and actual models of reasoning. In particular, we emphasize the classification model, an approach well suited to diagnostic and interpretive expert systems.

Chapter 2 introduces in more detail the class of problems for which expert systems are currently best suited: the classification problems. It reviews previous approaches to decision-making and places the expert systems approach in the context of past work.

Chapter 3 combines a historical overview of the specialized expert systems with a brief overview of several general knowledge representation frameworks recently developed in this field.

Chapter 4 gives specific advice on the design of an expert system with examples using the EXPERT design tool.

Chapter 5 describes several applications with specific prototype classes of expert systems.

Chapter 6 reviews some techniques of empirical knowledge acquisition and refinement, crucial for the development of truly mature systems.

Chapter 7 looks into the future and examines the directions of current research and the implications of using expert systems for practical applications.

1.7 Bibliographical and Historical Remarks

Nilsson has written an excellent general reference on the principles of artificial intelligence (Nilsson 1980). A comprehensive review of artificial intelligence systems may be found in a series of books edited by Barr and Feigenbaum (1981). There are numerous papers on the classification expert systems emphasized in this book. EXPERT was developed at Rutgers University and was first described in Weiss and Kulikowski (1979). EMYCIN is based on the original work done by Shortliffe and other researchers at Stanford University. EMYCIN is the subject of Van Melle's doctoral dissertation and one of the first papers appeared in Van Melle (1979). The PROSPECTOR system was developed at SRI International and is described in Duda, Gaschnig, and Hart (1979). Some thoughts on "knowledge engineering" for expert systems are described in Feigenbaum (1977). A brief review of practical results in expert systems is given in Duda and Shortliffe (1983). Other reviews of expert systems research can be found in Nau (1983) and Buchanan (1982). The views of a relatively large group of researchers attending a 1980 workshop on building expert systems have been synthesized in a book edited by Hayes-Roth, Waterman, and Lenat (1983).

2

Reasoning Methods
for Expert Systems

2.1 The Classification Model

Much of the power of an expert system comes from properly applying good reasoning techniques to a large store of problem-specific knowledge. The reasoning methods cannot be completely independent of the class of problems they must solve. Yet, at the same time, the reasoning procedures must not be so specialized to a particular problem that they are inapplicable elsewhere. In investigating expert reasoning we will look for those methods which are powerful, yet are generally applicable in describing and solving a broad class of useful problems.

The builders of expert systems have, over the past several years, developed a number of techniques for reasoning that are particularly well suited to the kind of knowledge used for solving problems in an expert fashion. A very natural way of expressing inferential knowledge is in the form of rules such as:

If: A is true and
        B is true and
        C is false
Then: Conclude X.

This is a type of production rule which has the general form:

If: Logical conditions are satisfied
Then: *Take the indicated action.*

There are a number of possible formalisms for describing these logical conditions and actions. When we try to assemble a system based on a collection of production rules, we have a *production system*. A production system must be able to evaluate the productions rules in the correct order, take the required actions, resolve conflicts, and request additional data which may be helpful in further processing of the production rules. These issues will be addressed in greater detail later in this chapter.

While much of our knowledge about decision-making can be expressed in rules of this form, there are often other kinds of information that human experts tell us are important in reasoning: knowledge of causal, temporal and functional relations among evidence, hypotheses, or parameters in various models that they may use in solving a problem. This knowledge typically underlies and supports the inferential knowledge expressed in rules. Yet most of the details of human expertise tend not to be based on well-understood theories, but are rather informal rules-of-thumb, or heuristics, for applying different types of knowledge to practical situations. When the heuristics are expressed as rules of the form shown above, we must consider the ways in which rules can be combined and compared to one another so as to come up with logically consistent solutions. And, depending on the kind of knowledge contained within the rules, we may choose to reason deductively or inductively, with logical certainty or with uncertainty, and with or without explicit representation of the detailed structure that makes up our data and conclusions.

The classification model is a natural framework in which production rules can be used. A classification system's task is to select a conclusion from a prespecified list of possibilities. In the abstract this implies three separate lists:

1. a list of possible conclusions
2. a list of possible observations
3. a list of rules relating observations to conclusions.

Although it is somewhat easier to follow the classification model using these three components, some implementations may combine the list of conclusions and the list of observations into a single list.

In this chapter we review some of the reasoning methods which the designer of an expert system might choose to use. We distinguish between the *knowledge content*, or *domain-dependent knowledge*, and *the structure* of the knowledge that is adopted for representation in the computer. The usefulness of an expert system representation and its associated reasoning processes comes from the constraints that they impose on the builder, forcing the inclusion of all knowledge that is important for arriving at conclusions, while imposing a discipline that will tend to exclude those elements that are less important for the decision making. Most existing expert systems provide a high-level language and related systems design philosophy that do not explicitly state how they constrain the builder of an expert system. In this book we hope to clarify the scope of what can be done in building models using the

EXPERT scheme, and to a lesser extent other methods for interpretation/classification problems. Putting the reasoning schemes into the perspective of existing formal methods of decision making is an important starting point.

To illustrate the various types of knowledge content that may be available about an expert problem, we will use a series of examples from a hypothetical problem of diagnosing a car which fails to start. This problem is quite similar to medical consultation problems, and the classification model has an excellent fit to this type of diagnostic problem. Although the examples are drawn from this specialized domain, the issues are universal. Each example has been chosen to deal with progressively more complex issues, but in this chapter we deliberately present the facts as they might be given to us by the expert, not worrying for the time being about how we might represent them in the computer.

2.2 General Statement of the Problem

Suppose we wish to design an expert consultation program that will give advice on the diagnosis and treatment of a car which won't start. The system would interpret data from various signs, symptoms and test results gathered about a malfunctioning automobile. How we go about building an expert system for this problem depends on how we decide to structure the problem, and how we manage to fit the human expert's knowledge into our structure. So, first we must try to see how the overall problem faced by the expert is broken down into subproblems, and how these interact. Then we must decide on a particular choice of representation for the expert's knowledge that will allow us to use more or less powerful methods of reasoning.

2.2.1 PROBLEM DEFINITION: REASONING SUBTASKS

In breaking down the kinds of reasoning involved in the consultation problem, we can identify several kinds of different reasoning subproblems:

1. given some pattern of symptoms, signs and test results, infer the possible diagnoses and the degree of certainty we can ascribe to them for the available evidence
2. given this same pattern of data, what other findings should be obtained if we wish to improve our certainty about the diagnosis
3. if we have tentatively established a diagnosis, how do we account for discrepant (unexpected or incompatible) findings
4. when and how do we move from tentative diagnoses to more conclusive ones

5. given that we have established a diagnosis, how do we proceed to choose a treatment, and how is a treatment to be chosen in urgent situations without having to wait for a diagnosis
6. for any of the above types of reasoning, how do we explain and support the conclusions reached or the recommendations suggested.

While there are other additional reasoning steps we can think of, for simplicity we will restrict ourselves to the above. In Figure 2.1 we have flowcharted the information-flow paths in a consultation system that shows how the reasoning can go back and forth between the different subtasks. The exact path taken in a specific consultation will depend on the nature of the specific evidence presented, and on what the expert expects to achieve through interpretation. Some experts might proceed very systematically through the stages, while others might wish to provide explanatory asides and weigh the information differently in repeated re-evaluations. Depending on circumstances, an expert may be able to take the time and expense to follow a particularly difficult path of analysis, while at other times, under pressure, he may opt for more direct lines of reasoning. So, in Figure 2.1 the user might omit some phases of explanation, with the reasoning moving directly from a preliminary to a final interpretation. This might be possible in reasoning about a straightforward set of alternatives with little uncertainty or ambiguity.

It is the task of the builder of an expert system to provide the representational flexibility that will enable a human expert to represent the various alternative approaches by which he can solve the problem.

2.2.2 PROBLEM DEFINITION: KNOWLEDGE STRUCTURE AND COMPONENTS

If we accept production rules as one of the most natural ways of structuring knowledge about the reasoning subtasks and specific inferences in a consultation, we must next ask ourselves: what are the knowledge components that enter into these rules, and how ought they be structured?

From a logical point-of-view, an inference rule of the If-Then type can be seen as an *implication* relationship which has two parts: the *antecedent* condition, or If part, which when satisfied (logically found to be true) leads to an assertion that the consequent, or Then part, must be true also.

In our consultation example, we might have a rule of the form:

If: The car won't start
Then: *Consider the possibility of electrical system problems.*

Such a rule says nothing about the inverse logical condition: if the consequent (the car is considered to have electrical system problems) is known to

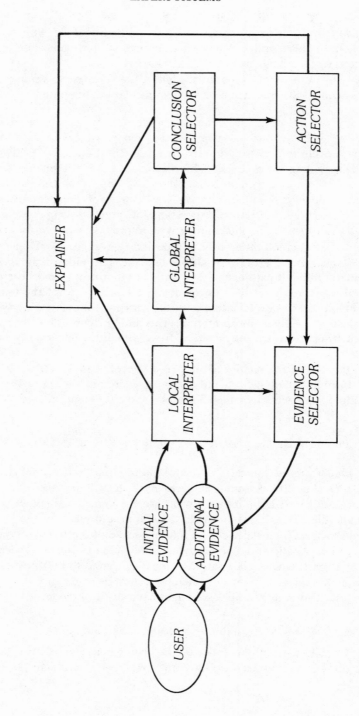

Figure 2.1:   Information flow for reasoning subtasks in an expert system

be true, the antecedent condition that the car won't start might be true, but then again, it might just as well be false.

The antecedents and consequents are, logically speaking, *propositions*. That is, they are assertions that some fact is true, or else that some hypothesis or conclusion is true. As an antecedent, there is nothing to stop us from including a combination of assertions that must all be true together in order for the consequent to hold. An example was the first rule given in this chapter. Similarly the consequent could represent some conjunction of propositions, as in the following example.

If: The car won't start and
          the fuel gauge reads empty
Then: *The car has no fuel and*
          fuel should be placed in the gas tank.

On the other hand, many different rules could be written with the same consequent, but with different antecedents. Together they would cover the alternative conditions under which one could infer the consequent. For example, other rules very similar to the one shown above could have alternatives to the gas gauge reading, such as the number of miles traveled since fuel was purchased.

If there are only a few such alternatives, we might wish to combine them as a single disjunctive antecedent in a single rule, but if there are many, it will be more convenient and logically equivalent to write a set of separate rules. For example,

If: The car has either of these symptoms:
          the starter doesn't crank or
          the starter makes odd (grinding) noises
Then: *Consider the possibility of a starter malfunction.*

could be written as two rules, each with the same consequent and each with a single one of the conditions listed above.

From an *expert systems* point-of-view, what kind of propositions are likely to be encountered as consequents and antecedents of our rules? If we think in terms of the production rules shown above, a simple and natural answer might be: antecedents would be patterns of data; consequents would be the conclusions that we infer from these patterns, such as diagnoses, or treatment recommendations.

These, of course, are not the only possibilities—an antecedent could be a diagnosis and the consequent a treatment:

If: The carburetor is flooded
Then: *Depress accelerator to floor while starting.*

For explanations, antecedents would be the paths of inference, and consequents would be the reasons for these inferences in terms of the goals satisfied or the underlying knowledge that justifies them. For example:

If: A conclusion of battery discharged has been "tentatively" reached and the reasoning shows that the following rules were invoked in reaching the conclusion:

- Rule 106: using "car won't start" as an antecedent
- Rule 139: using "headlights dim" as an antecedent

Then: *The conclusion has still not yet been confirmed by a battery-fluid tester.*

Here we see how the explanatory conclusion embodies the underlying assumption that for this problem an objective test result is essential if the diagnosis is to be "definitely" confirmed.

In medical situations, the patterns of symptoms, signs and tests lead us to infer that a particular diagnosis is more likely than some other. A final diagnostic interpretation is usually not reached in a single step by directly inferring the diagnosis from the data. More frequently, intermediate hypotheses about possible causative factors, syndromes, or plausible mechanisms of dysfunction will be used by the physician as stepping stones in the process of arriving at a more definitive conclusion. These *intermediate hypotheses* are an important component of medical reasoning, which it shares with other types of sophisticated expert interpretation, whether it be oil well-log analysis or high technology equipment diagnosis.

Therefore, if we wish to build an expert system we must usually find out from the human expert which important intermediate reasoning hypotheses are needed to build up coherent decisions in that specialty. In our expert system, then, we will also need to consider patterns of these intermediate hypotheses as possible antecedents of our rules. For example:

If: The car has a battery discharge problem and
the battery has just been replaced

Then: *Consider testing the charging system (alternator or regulator).*

Yet another type of reasoning rule is frequently used in expert reasoning: constraints among the observations. If, for instance, we know that the person examining the car is a skilled mechanic, we may expect more heavy reliance on specific testing instruments to form an interpretation. If on the other hand, our car examiner is a novice, we may not expect results from tests that require instrumentation. Thus, conditions of testing and observation are likely to be antecedents that will assert that other observations can be expected to be either present or absent. For expediency, we can also treat some very well known facts as observational constraints: knowing that a car has electronic ignition should exclude any need to inquire about points. Although there is a quite detailed mechanical and electrical model from which such inferences could be generated, it would hardly demonstrate human expertise to do so: we describe such inferences as well-established knowledge that can be directly encoded in rules. In many other expert reasoning problems, similar but less obvious alternatives for representation arise, but it is useful to be able to encode relations among observed facts directly.

A different type of reasoning rule best captures the information that will help support or confirm a tentative interpretation. The consequent of such a rule must be of a special type: it is the expectation that a particular test result, if found to be true, will then improve our confidence in the conclusion. For example:

If: The car won't start and the starter cranks slowly and
         the battery has not been tested
Then: *A battery test should be performed.*

This is just simply a special type of conclusion which allows us to make decisions in a stepwise fashion. If the test is subsequently performed and the expected result is not observed, we might wish to either redirect our conclusion or check for possible errors or inconsistencies in our data.

When our conclusion is to recommend a treatment, we are doing so expecting to improve conditions for the better. If this doesn't happen, we must then try to assess why the desired effect didn't occur, whether because our recommendation was not carried out, or whether it was faulty because of limitations in our model or because the diagnoses or conclusions on which it was based were incorrect.

To summarize, the propositions in an expert system's rules are likely to include factual statements about observables (findings) and hypotheses (conclusions) at different levels of complexity and uncertainty. The rules are likely to link findings to other findings, patterns of findings to hypotheses, and patterns of hypotheses to other hypotheses. In developing an expert system, we must examine the scope of relations that might be given to us by an expert. This brings us to the specification of different levels of knowledge content for the system.

2.3   Definition of Knowledge Elements: Conclusions and Observations

We have looked at only a few of the problems which might cause a car not to start, such as battery or starter problems. One of the first questions faced by the builder of an expert system is: does this adequately cover the spectrum of possibilities? Talking with our human expert we will rapidly find out that the simplification has been drastic indeed. If we start initially with only a few possible diagnoses, we quickly learn that to build a system which comes close to the human expert in performance we will need many diagnostic conclusions for each of our initial ones. But then there is the issue of relations among conclusions. When we begin with a few simple ones, they can be chosen as mutually exclusive alternatives. Once the scope of conclusions is enlarged this simple situation usually doesn't apply: a car may have both battery and starter problems simultaneously.

Finally there is the question of uncertainty in the conclusions: with relatively little but suggestive initial data, our diagnostician might be willing to state that a specific diagnosis is possibly or probably present, while not being definite about it. And what about predictive statements of risk even if there is no evidence of current dysfunction? How should we define uncertainties for them? The various formal methods of reasoning provide some answers to these questions.

Comparable concerns arise in defining the observations more carefully. How are they to be qualified in detail, and should they have an attribute of strength? In medicine, symptoms are often qualified to denote mild, moderate and severe manifestations or confidence in the presence of disease (e.g. possible, probable or definite). Relationships among symptoms are important, both at the observational level (Is one easier to measure than another?) and at the conceptual level (Are certain symptoms impossible in the presence of others?).

Answers to many of the above questions will have to wait for the specific examples below in which we introduce the appropriate degree of structure, complexity and knowledge in the problem formulation. For now we will consider that we have defined a simple *list* of conclusions (which, for simplicity, are taken to be mutually exclusive in the present example), and a *list* of observations, any of which may be applicable to a problem.

Just knowing the relevant evidence and conclusions is not sufficient for reasoning to proceed. The next level of knowledge content considers the simplest kinds of inference links (which we must express as rules), through which conclusions can be reached.

2.4   Logical Association of Evidence and Conclusions

If our human expert wants to give us the simplest possible information relating observations to conclusions, he can provide logical associations between the two. In English these can be expressed in a variety of ways. For example:

1. car won't start is associated with electrical system problems
2. fuel gauge reads empty is indicative of an empty fuel tank
3. a typical picture for fuel system problems includes: strong gasoline odor, car won't start, normal starter cranking.

All these statements might on the surface appear to be of the same type, but if analyzed more carefully they fall into several types.

One way of grouping the statements into types is to consider the nature of the inferences that they state. There are three major types according to whether they express an inductive inference, a deductive inference, or an unspecified association.

Other groupings could be made, depending on whether single or multiple pieces of evidence are used and whether the conclusions are to be arrived at with full confidence or with some degree of uncertainty. These properties come from a closer study of the logical and semantic links present in the statements. In this section we now contrast the three different types of inferences that are specified by these three statements.

2.4.1 INFERENCES ABOUT CONCLUSIONS
FROM THE EVIDENCE

When evidence points to a conclusion which is only one conclusion among many possible conclusions for the problem, it has the effect of making us more confident that the conclusion is in fact reasonable or true. In logical terms we are carrying out an inductive reasoning step, inferring a general law or hypothesis as being applicable to our problem (e.g. conclude the patient has a specific disease) from the specific data or evidence presented. Statement 2 can be paraphrased as follows:

If: Evidence E is found to be present
Then: *Conclusion C can be inferred or rejected.*

This is the type of association immediately useful in arriving at a conclusion. But if all associations are of this form, with no indication as to which item of evidence is more important than which other, it becomes difficult to assess the cumulative effect of applying many such inference associations. How one should handle consistently and together a large variety of statements of this kind is a question that requires deeper insight into the structure of domain knowledge that supports the inferences. Nevertheless, if we are limited to the above kind of association alone, we can still carry out a simple form of reasoning: we can list all the conclusions that are pointed to by the evidence in each association.

2.4.2 INFERENCES ABOUT EVIDENCE
BASED ON A KNOWN CONCLUSION

Statement 3 is a description of a *typical* set of observations associated with a conclusion. Logically they can be paraphrased as:

If: The conclusion is C
Then: *Evidence E is likely to be present.*

In medicine, this kind of statement comes from clinical studies where a group of patients with well-established diagnoses are studied and found to have certain typical observations. These inferences are not directly useful in decision-making because they don't tell us how to make an inference, but just what to expect if we already have made it. If our evidence happens to match the *typical* case of conclusion C we might feel inclined to infer that C is indeed the conclusion to make, but we have in fact no way of knowing whether this is right or not, unless we know that no other conclusion can also have the same typical pattern that C has. Thus we need more information than simple associations to make use of this kind of link between conclusions and evidence.

But there is a special variant of this kind of link that can be useful in decision-making. If the conclusion is not really known, but just one of several plausible alternatives, rules that give us the patterns of evidence that are likely to be present if the conclusion were true might be used to suggest searching for just those patterns, since it is likely that they will contain elements that are also useful in inferring the diagnoses. So while no conclusive inferences may arise from them directly, they can help to focus the accumulation of evidence.

2.4.3   UNSPECIFIED TYPES OF ASSOCIATIONS BETWEEN CONCLUSIONS AND EVIDENCE

Statement 1 is of an unspecified type. It merely tells us about the association, but gives no indication of whether evidence can be deduced or inferred from conclusions or vice versa. If presented with this association we might at best say: the conclusion is a possibility, but we cannot really say for sure, because it might just as well be an alternative conclusion. If we are given a lot of information of this kind in a knowledge base, it will clearly not be easy to distill much that is useful.

2.5   Explicit Statements of Uncertainty about the Conclusions

For the time being, let us forget about various forms of production rules which relate evidence to conclusions. The simplest procedure for reaching a conclusion based on evidence would be to have stored in a data base all possible patterns of evidence together with the conclusions associated with them. For any given case, one would simply look up the the conclusion for the specified pattern. This is clearly optimal if there is only one conclusion attached to each pattern of evidence; we would always get the right answer. In terms of production rules, the antecedents would consist of all possible combinations of evidence.

If we notice that a pattern of evidence is not unique to a single conclusion, then we will need to assign probabilities to patterns in the data base indicating the likelihood of a particular conclusion given a specific pattern. Under these conditions, there are statistical procedures which are theoretically optimal. That is, if the *true* population statistics from which our data are sampled are known exactly, they will allow one to make decisions so that a minimum rate of error is guaranteed for the population as a whole.

As one suspects, however, the data base lookup approach, even without probabilities or numerically valued evidence (such as weights), will lead to a combinatorial explosion of possibilities. If we have 1000 pieces of evidence which are either true or false, then there are 2^{1000} possible combinations. From a practical point of view, we will need to find procedures which reduce the dimensionality of our task. The notion of production rules and expert systems is one such heuristic approach to this problem. It is important, however, to be aware of procedures which are theoretically optimal and their relationship to the practical world, so that we may make intelligent design choices in building an expert system.

2.5.1 PROBABILISTIC STATEMENTS

By using probabilities we make sure that we can compare the results of our analysis with a frequency interpretation. This has an empirical attraction which should not be underestimated, but there are several reasons why pure probabilistic analyses have not been very popular in expert systems. These will be discussed in detail later. On first reading, those readers who do not like mathematics may wish to skip to section 2.5.3.

2.5.1.1 BAYES' POSTERIOR PROBABILITY METHOD

This is probably intuitively the most appealing approach from among the statistical theories. If our conclusion is to be selected from a set of mutually exclusive alternatives (for simplicity we consider only two), the Bayesian approach in its most direct form tells us to choose the conclusion with the highest probability for the given data. In terms of our examples, we might consider asking the probability of a malfunctioning starter for a car which shows a symptom suggesting this problem. The kind of knowledge we would like the expert to give us is:

If the starter is making odd noises the probability of a bad starter is approximately 75%.

The hidden implication of such a statement is that the probability of the opposite condition must be 100% minus 75%, or 25%. In effect, there is a companion rule implied by the above which can be stated as:

If the starter is making odd noises the probability of a good starter is approximately 25%.

Many studies have shown that experts do not easily carry probabilities such as the above in mind, and even when they do report them, the numbers do not turn out to be true probabilities, because the reporting specialist will often deny agreement with the hidden implicit probability (25% in this case) of the opposite conclusion.

If an expert cannot provide subjective estimates of the Bayes' probabilities, they can be computed indirectly from the direct or conditional probabilities which are derived from empirical studies, such as:

If a car has a bad starter then the starter will make odd noises in 87% of the cases.

To convert this probability into an inductive or Bayesian one, we need one more controversial ingredient: the prior probability. This would come from a statement of the kind:

The probability of a bad starter (when a car won't start) is 2% (before looking at any specific symptoms).

How was the 2% arrived at? Is it the percentage of all cars in a repair shop that have starter problems as carried on our records today, or over the past year, or over the past ten years? Why use a specific repair shop as a base for counting; why not the number of starters sold versus other parts? Should we distinguish cars by models or years to get these figures? These and many other questions make it difficult to decide which number to use.

The theory, however, tells us that once we have these numbers, we can derive the probability of class C, given evidence e, by Bayes' formula:

$$P(C \mid e) = P(e \mid C) \cdot P(C)/P(e) \tag{2.1}$$

where $P(C)$ is the prior probability and $P(e)$ is the unconditional probability of the evidence:

$$P(e) = P(e \mid C) \cdot P(C) + P(e \mid \sim C) \cdot P(\sim C) \tag{2.2}$$

Mathematically this is quite simple once the argument over the source of the numbers has been settled. In the example given above, $P(C) = .02$, $P(\sim C) = .98$, $P(e \mid C) = .87$, and we see that we now need a new direct probability $P(e \mid \sim C)$, which would require having collected statistics in our study which then provided a statement of the type:

The probability of a normal starter making odd noises is 7%.

With this in hand we can now compute P(Bad starter | odd noises)

$$.87 \cdot .02/(.87 \cdot .02 + .07 \cdot .98) = 20.2\% \tag{2.3}$$

which is considerably lower than our expert's estimate of 75% given in the first statement of this section! Why the discrepancy? It is difficult to know unless the expert is able and willing to provide his assumptions in probabilistic terms. The most obvious reason for discrepancies, however, would be a different assumption of prior probabilities implicit in the expert's choice of the Bayesian values for $P(C|e)$. We might even try to guess at these values by working backwards from his numbers and the ones for the direct probabilities $P(e|C)$.

In this special case where we consider only two decisions at a time, such as having a bad starter or not, the probability of the conclusion $P(C)$ is related to the probability of its negation as $P(\sim C) = 1 - P(C)$. Then we can relate $P(C)$ to the other probabilities by the formula:

$$P(C) = 1/\{[P(e|C)/P(e|\sim C)] \cdot [1/P(C|e) - 1] + 1\} \qquad (2.4)$$

which can be derived from Bayes' formula. For our example, if we were to take the 75% of the expert as the correct Bayes' probability, we would find that in order for it to be consistent with the direct probabilities, the expert's prior probability would have to be:

$$P(C) = 1/\{[.87/.07] \cdot [1/.75 - 1] + 1\} = .19 \qquad (2.5)$$

which is more than nine times larger than the prevalence figure of 2% given to us. Thus, for whatever reason, the expert is using a very different estimate of the priors in his subjective probabilities, if indeed he is using probabilities at all.

Another factor that can always be at fault is the validity of the direct probabilities. Any collection of statistics is gathered under circumstances difficult to replicate in all other environments. If the test results used for decision making are strongly affected by some local condition (method of measurement, bias introduced in observation, etc.), then the statistics will not be transferable to other sites, and their applicability for developing decision rules will be suspect.

In the equation for Bayes' rule (2.1), we are trying to evaluate conclusion C given evidence e. Evidence e is usually not a single observation, but instead a combination of many pieces of evidence, i.e. it needs a joint probability. In fact, this probability is exactly the type of probability referred to in section 2.5; it is one of a very large number of possibilities, which must be looked up in a data base for the exact combination of evidence. Therefore, to use Bayes' rule for optimal decision-making we would need a very large data base of cases indeed. Because this formula can not be used exactly in the real world, people have used various approximations or simplified assumptions. The most common is the assumption of conditional independence, which says that, for any two observations e_1 and e_2, $P(e_1|C)$ and $P(e_2|C)$ are probabilistically independent. While this may give reasonable answers for small problems, most complex problems have dependencies among their data, making this simplification inaccurate. With the assumption of independence, the following formula holds:

$$P(C \mid e) = \prod_i P(e_i \mid C) \cdot P(C)/P(e)$$

where:

$$P(e) = \prod_i P(e_i \mid C) \cdot P(C) + \prod_i P(e_i \mid \sim C) \cdot P(\sim C)$$

Bayesian decision-making becomes considerably more complex than described above if we are concerned not only with the effect of the prior probabilities, but also with differences in the cost that we associate with each type of possible mistake. This extension of the original Bayesian theory will be discussed under the heading of *decision theory*.

2.5.1.2 HYPOTHESIS TESTING

Because of the difficulties in choosing consistent prior probabilities as shown above, the statisticians Neyman and Pearson in the 1930s proposed a different approach to statistical inference which relies only on the direct probabilities and some choice of performance criterion. This method, called hypothesis testing, was applied to signal detection and recognition work in the 1940s, and extended to a variety of decision-making problems later. The basic idea is to assume that because of the uncertainties in our data we must expect to arrive at mistaken conclusions some of the time, but we want to design a decision rule in such a way that we limit these errors to a small, and hopefully tolerable, fraction of the decisions that we make.

In simple two-category decision problems (conclusion C versus the opposite conclusion $\sim C$), we can expect two types of errors to arise. One is the mistake of deciding C when in fact $\sim C$ is the case, and the reverse is to have decided $\sim C$ when C is true. In medicine, a decision for (diagnosis of) disease is termed a "positive," whereas one for health is called a "negative." Thus a "false positive" decision would be one that errs on the side of diagnosing disease when a patient is really healthy, and a "false negative" would be a diagnosis of health when the patient is in fact sick. Traditionally physicians have been much more concerned about missing sick patients, so the false negative mistakes are the ones that must be controlled most carefully. Likewise, oil exploration specialists are often more concerned about missing a large oil find than in drilling a single dry hole. There is a tradeoff between the two kinds of errors for any choice of decision rule based on a fixed set of data. The method of hypothesis testing ensures that the items of data will be combined in the best possible way to guarantee that the error of one kind (such as the false positives) is minimized subject to a constraint on the errors of the other kind (false negatives in this case). Because of the mathematical constraints we cannot minimize both errors simultaneously.

As an example, consider the problem of a car not warming up properly because of a faulty or improper thermostat in the car's cooling system. The thermostat opens fully at an abnornomally low temperature; thus the car runs at an abnormally low operating temperature. Consider a single test, t_1, such as the temperature gauge reading (i.e. the coolant temperature), which has probability density distributions shown in Figure 2.2. We also show the resulting errors FP (false positive) and FN (false negative) for the particular choice of decision threshold of D_1 (e.g. 210°F) given in the subjective decision rule. As the decision threshold is shifted to D_2 (e.g. 225°F) we see the change in the error rates, with the false negatives becoming fewer and fewer as the threshold is increased, thereby diagnosing more and more thermostats as malfunctioning. In the limit, we reach the best case as far as false negatives are concerned (none) at the cost of 100% false positives! That is, we call all thermostats broken regardless of the value of the test. We can carry our shifting of the threshold to the opposite extreme and obtain an equally ridiculous case of no false positives at the expense of 100% false negatives. What makes sense is a choice of decision threshold at some point between the two extremes. The actual choice will depend on the percentage of false negatives that we are willing to tolerate. For instance, if we are at worst willing to accept a small number of false negatives (e.g. 8%), the corresponding decision threshold for t_1 will give us an increased false positive rate (e.g. 30%).

Suppose that we had a better test to detect a faulty thermostat. In this test, t_2, we remove the thermostat from the car and put it in a container of coolant and heat the coolant, observing the temperature at which the thermostat opens. The probability density distribution for this test, t_2, is shown in Figure 2.3. We can see that because there is less overlap between the bad thermostat and good thermostat groups for this test, we can get a much better tradeoff (e.g. 5% FN rate might result in 10% false positives.)

The tradeoff between false positives and false negatives can be plotted on a graph as shown in Figure 2.4, for both test t_1, and test t_2. This makes it easy to compare the tests in terms of their decision-making tradeoff characteristics. We see immediately that for most reasonable choices of a tolerable FN rate (small values), we can get better performance out of t_2 than t_1.

The mathematical formulation of the hypothesis-testing decision scheme involves taking the ratio of the probability density value for the data point under consideration and comparing it to a threshold, which is a constant that depends on the choice of a tolerable error rate (either FN or FP). As a formula this is expressed as:

$$f(e \mid C)/f(e \mid \sim C) > t(a) \text{ decide C as conclusion}$$

whereas if

$$f(e \mid C)/f(e \mid \sim C) < t(a) \text{ decide} \sim C \text{ as the conclusion.} \qquad (2.6)$$

Here, the quantity a can be the FN error rate or FP error rate as desired.

In situations where the evidence consists of discrete-valued variables, such as answers to yes/no questions, we must substitute a probability value for the density value in the equations above.

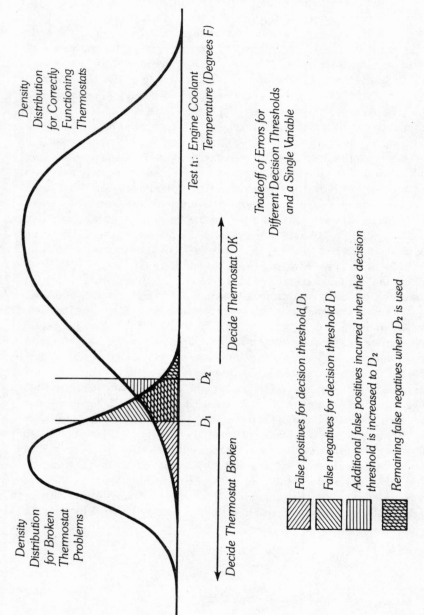

Figure 2.2:   Error rates derived from a decision rule based on the coolant temperature for diagnosing a faulty thermostat

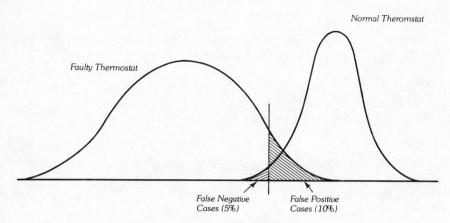

Normal Theromstat

Faulty Thermostat

*False Negative
Cases (5%)*

*False Positive
Cases (10%)*

*Test t₂: Temperature of coolant when thermostat is
observed to open completely.*

Figure 2.3:   Example of a test for which a better decision rule can be written

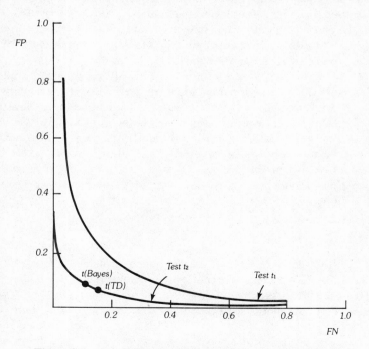

Figure 2.4:   Tradeoff between false positives and false negatives

There is a direct relation to the Bayesian decision-making that then becomes apparent:

$$P(C \mid e)/P(\sim C \mid e) = 1 \qquad (2.7)$$

is the Bayesian decision equation for the boundary between the two decisions, and it can be transformed into the equation:

$$P(e \mid C)/P(e \mid \sim C) = P(\sim C)/P(C) \qquad (2.8)$$

by use of Bayes' formula.

In the hypothesis testing situation, the equation for the boundary is:

$$P(e \mid C)/P(e \mid \sim C) = t(a) \qquad (2.9)$$

which shows that the Bayesian case can be viewed as a special case of the hypothesis-testing decision rule with the threshold set to the value $P(\sim C)/P(C)$. This is illustrated on the tradeoff curve in Figure 2.4 as point t (Bayes).

From a theoretical point of view the goodness of either the Bayesian or the hypothesis-testing approach to decision-making depends on having the ideal or population statistics that *truly* represent the type of variability in the tests taken over the groups we are trying to distinguish. If we use sample-derived densities rather than the infinite sample-population ideal, we are likely to make mistakes, and the method is no longer guaranteed to be truly optimal in giving the best tradeoff. However, we can use large sample statistics to approximate the population densities and do well with this approach.

The real usefulness, but also the source of difficulty in practical application of this method, becomes apparent when we deal with decisions involving more than a single test. If we wish to see whether combining several test results will give us better decision results than taking one at a time, we need the joint distribution densities which are difficult to accurately obtain from samples. For this reason people often resort to simplifications, such as straight line separations between the groups. These are called linear discriminant functions and are a popular method that involves heuristic elements with a statistical basis, as will be shown in section 2.5.2.1.

2.5.1.3 DECISION THEORY

When we wish to take into account different costs of making a false positive versus a false negative mistake, we can apply methods of decision theory that are an extension of the Bayes' method.

Suppose we are told that false negatives are ten times worse than false positives, and we know that the prevalences of C versus $\sim C$ (e.g. disease versus non-disease) are 0.66 and 0.33 respectively. We can then write the average cost of making a mistake as:

$$\text{Cost} = \text{Cost(FP)} \cdot P(\text{FP}) \cdot P(\sim C) + \text{Cost(FN)} \cdot P(\text{FN}) \cdot P(C) \qquad (2.10)$$

Here the probabilities of FP and FN are computed from the decision rule application.

To minimize the above quantity, the decision rule that is best turns out to be nothing but another variant of the likelihood ratio comparison test, but with a different threshold: Decide C if

$$P(e \mid C)/P(e \mid \sim C) > P(\sim C) \cdot \text{Cost(FP)}/P(C) \cdot \text{Cost(FN)} \qquad (2.11)$$

with the inequality reversed corresponding to the $\sim C$ decision. The same kinds of problems that were discussed with the Bayesian method are applicable here. In addition we have to worry about the subjective nature and choice of the relative costs. Just as with the Bayesian case, we see that we can consider the threshold derived here as a single point on the FP/FN tradeoff curve. This is shown in Figure 2.4 as $t(DT)$.

An advantage of using the decision-theory approach is that it gives a subjective justification for the choice of threshold in the hypothesis testing spectrum of options. Yet experts are usually quite hesitant about assigning relative costs to the two types of error, and arriving at reasonable decision thresholds can be quite difficult.

2.5.2 Approximate Statistical Methods

In this category we find several approaches which are useful when we do not have many examples with good endpoints from which statistics can be calculated. These methods are not optimal except under restricted conditions: they rely on general assumptions about the smoothness of data or decision boundaries and less on knowing the detailed distributions. The most popular decision method in this group is that of *linear discriminant functions*, followed closely by the *nearest neighbor classifier method*.

2.5.2.1 Linear Discriminant Functions

In general we cannot expect the probability densities to be smooth mathematical functions. When small samples are available and no good estimate of the densities is forthcoming, we suspect any irregularities in the sample density shape to represent primarily variations from the sampling process rather than true underlying distributional changes. It is often wise to then ignore the higher order fluctuations in the statistics that give rise to the irregularities of density shape, and look only at the low order statistics, since they are more reliable even for small samples.

The method of linear discriminant functions does the above in several ways, depending on the type of criterion that we seek to satisfy in our choice of decision method. The first and most popular method is the work of Fisher, and consists of developing a score upon which the decision is to be made, a weighted sum of the evidence or test results, e_i that enter into the decision. The weights are derived from a formula which makes the resulting scoring function the best one in separating the classes. Thus, we decide C if:

$$S = w_1 \cdot e_1 + w_2 \cdot e_2 + w_3 \cdot e_3 > \text{threshold} \qquad (2.12)$$

and decide $\sim$ C if S < threshold.

In their most elementary form, each of the weights is found by calculating the difference between the means of the variables and normalizing by dividing by the standard deviations. This gives comparable scaling to all the variables, and produces a mathematically tractable function.

2.5.2.2   NEAREST NEIGHBOR METHODS

If we don't even trust the mean and variance of our samples as reliable statistics, we can fall back on this method which uses individual samples with well-defined endpoints as the patterns against which new cases of unknown classification are to be compared. The approach is very simple: we decide for the class corresponding to the pattern that is the nearest neighbor, i.e. closest to matching the new case. For example, if class C and class $\sim$ C are characterized by the following binary samples of e_i, then a pattern 111 of unknown class would be classified as belonging to C, because it is closer to a sample of C than $\sim$ C.

$$C = \{110, 011\}$$

$$\sim C = \{000, 001\}$$

The attractions of the nearest neighbor method are great if we are willing to match our unknown cases against all cases with known endpoints, or if we can summarize them by representative cases. As more samples accumulate, it can be shown that the performance of the nearest neighbor method will be bounded by the optimal Bayesian error rates, giving confidence in the results.

Among the drawbacks of this approach are the usual dimensionality issues, and the lack of an accurate closeness measure that distinguishes the relative importance of recorded observations.

2.5.3   RELATIONSHIP OF UNCERTAINTY METHODS TO THE RULE-BASED REPRESENTATION

We have examined the statistical approaches because from a theoretical perspective they can be optimal. The practical application of these methods for real models is not feasible for models of even modest dimensions (10 classes and 100 observations), because of the dimensionality issues discussed earlier. Those relatively small systems built using the statistical schemes, therefore rely on some simplifying assumptions.

Because we cannot look up the answer in a database, we must find approaches which reduce dimensionality, i.e. reduce the numbers of combinations of evidence and of answers which must be considered. The concept of using production rules is quite simple. Why not go to an expert and have him tell us which combinations of evidence are the important ones for decision-making. In practice, many classification problems can be highly abstracted to identify a core of important logical combinations. Instead of relying on the use of samples of an expert's cases, we go directly to the expert and try to extract this information. Besides not relying on indirect information from cases, by going to the expert we can potentially get far more representative information, and information covering a much broader spectrum of alternatives than might have been possible with a limited number of cases.

The process of abstraction must go further than finding key combinations of evidence for a given conclusion. A production rule in an expert system will usually contain not only elementary observations which have been the focus of our attention in examining the statistical approaches, but also intermediate conclusions which abstract and summarize several observations. This will be of major importance in reducing the potential number of combinations.

Once we leave the realm of table lookup for the correct answer, we will be faced with patterns of evidence that are not mutually exclusive. We must have some means of choosing among the patterns when several are indicated, each with a different likelihood. There is no general way of knowing how to combine these likelihoods in a statistically correct manner, without knowing the exact and complete joint probability. Several heuristic approaches have evolved for this problem; each has its advantages and shortcomings. One of the important proposed solutions is less reliance on mathematical functions for combining patterns and more on descriptive means of finding the appropriate pattern. This and other problems will be addressed later.

Among the potentially important assets of the production rule approach is that it provides the means of understanding how a decision was reached and is able to *explain* and *correct* erroneous conclusions. This is quite difficult in the statistical approaches which, to some observers, appear to be formulas that are invoked for *extracting* a number which summarizes the results of a nonintuitive process.

Many of the practical problems which must be faced in designing an expert system involve the frequent lack of firm endpoints or conclusions. This is particularly true in the early stages of the design of a model, when we may not know the exact conclusions that we wish the system to reach, or the complete set of observations which are relevant to the problem. We must therefore have some flexible approach which allows us to experiment and vary our model without drastic consequences. If one is not relying on case data for driving the system, the production rule approach is highly flexible for modifying the knowledge base, and new rules are unlikely to create burdens on the system. Gathering data is difficult enough without having to cope with dynamic changes in the conclusions.

If we could solve the dimensionality problem, one might propose making humans conform better to the requirements of formal methods: ask them to never make a statement that didn't include a clear definition of the alternative hypotheses and their probability measures. While this would certainly solve our formal reasoning problems, we expect an expert system to work with the same degree of incompleteness as does human expert reasoning. From the formal point of view, we must be willing to work with partially complete knowledge bases.

Several of the issues of designing an expert system using production rules and the classification model will be discussed in Chapter 4. With key abstractions of knowledge using production rules, we will see that many problems can be solved with little need for significant use of strict likelihood measures. First, though, let's look briefly at the general issues addressed by rule-based systems.

2.6   Logic for Evaluating the Antecedents of Rules

Before examining *production systems*, let's look at the simple evaluation of a single production rule. Here we must ask how logically to evaluate the components in the rules; that is, the antecedents and consequents.

The antecedents are the ones that present the most interesting questions. As mentioned above, an antecedent is a logical proposition that can consist of an arbitrary number of subpropositions. Each of these represents either a fact that is taken to be true by direct observation (the evidence), or else a fact directly deduced from evidence, or a hypothesis that is inferred from evidence.

The logical evaluation of antecedents can then be viewed as the problem of deciding how to combine conjunctively (AND) or disjunctively (OR) the truth or confidence values in several propositions of the above type in order to make the consequent true.

2.6.1   BOOLEAN EVALUATION

The logical evaluation of premises with items that can only be true or false is easily carried out, so rules that only contain evidence (or facts directly deduced from evidence) on their left-hand side (LHS) antecedents (IF conditions) can be handled according to Boolean methods. If they contain a conjunction we require all items to be true for the entire LHS to be true; if a disjunction, a single true item will suffice to make the LHS true. In expert systems we also have to worry about hypotheses on the LHS, and this brings up the question of evaluating credibility measures and propagating them to the consequent (THEN conditions), or right-hand side (RHS) of a rule.

2.6.2 PROBABILISTIC EVALUATION

Each proposition that represents a hypothesis on the LHS of a rule must have a truth value associated with it during evaluation. If we allow for true probabilistic statements, there is no universal way of combining probabilities attached to individual propositions so as to derive a probability for the conjunction of the propositions. Either the joint probability of occurrence of the conjunction is known a priori or else we can assume partial or complete independence and multiply the probabilities—which is an approximation unless we are sure that the propositions are truly independent. The probability of a disjunction is simply the sum of the probabilities of the individual propositions minus the excess due to included conjunctions. Thus disjunction also requires knowing or approximating joint probabilities for conjunctions.

2.6.3 CONFIDENCE FACTORS

Most heuristic methods have sought to justify their approach by some quasi-probabilistic interpretation. Yet there are other logical formalisms within which the reasoning of experts can be described. One such formalism is *fuzzy logic*, which uses a multi-valued *membership function* to denote membership of an object in a class rather than the classical binary *true* or *false* values of Boolean logic. Thus, a person will not be considered to be either old or young, but, depending on his actual age, can have certain degree of being old, and a different degree of being young, both of which will depend on the membership functions that people define for the concepts of *old* and *young*. If we take a fuzzy logic approach, we handle conjunctions by using the minimum weight over the constituent propositions, and disjunctions by taking the maximum weight to attach to the whole LHS.

The most common representation of heuristic weights, or *confidence factors*, is to use numbers greater than 0 for positive evidence and numbers less than 0 for negative evidence, e.g. numbers between -1 and 1. These numbers are used solely as heuristics, and no criterion of optimality is associated with them. Many expert system designers believe, however, that this distinction between positive and negative evidence is helpful in extracting knowledge from the expert.

Various heuristic weights schemes have been proposed, such as a confidence measure used in the MYCIN system. Some approaches to writing production rules (EXPERT) rely on a prior estimate of the weight attached to conjunctions. Others (MYCIN, PROSPECTOR) use a fuzzy logic approach, the minimum confidence for a proposition in the premise. Where the heuristic methods differ is in the evaluation of disjunctions, for example when several production rules are satisfied for a single conclusion. A variety of heuristic scoring functions for combining confidence measures have been proposed, all making assumptions that work well under certain conditions, but also have serious drawbacks. The simplest scheme for disjunctions is the fuzzy logic approach of taking the maximum weight. This is the approach we have taken in EXPERT. A more detailed discussion of using confidence factors in an application system will be given in Chapter 4.

2.7   Logic for Evaluating a Set or Sequence of Rules

Here we consider how to combine several rules so that we obtain smooth and accurate reasoning. We must also keep in mind that human experts are often able to specify sequences of decisions that are related to their experience with prototype cases.

2.7.1   DECISION SEQUENCES, DECISION TREES AND INFERENCE NETWORKS

When the expert codifies his decision processes as a sequence of steps, he usually starts with evidence that might present itself, and then asks the question: if I have this evidence, what conclusion(s) does it point to, and what other evidence do I need to gather in order to increase my certainty in one of the conclusions, and reduce my certainty in all the alternatives? If he then proceeds to gather the additional piece of evidence that is most likely to discriminate between the interpretive alternatives, he will then take a further step, depending on the exact nature of the results. If enough information has been provided for him to make a definitive conclusion and exclude the alternatives, then the decision process terminates at that step. Otherwise, he will proceed to ask for additional evidence as many times as needed to arrive at the final decision point. In the "car won't start" example, an observation of slow cranking will direct the questioner to the possibility of a weak battery, whereas normal cranking will raise other possibilities. The subsequent line of questioning will be different depending on this single outcome, and on other subsequent outcomes.

This suggests that in tracing the logic of an expert step-by-step, the builder of an expert system would have to fan out from each possible starting point to a different sequence of steps in order to cover all the possible alternatives of reasoning. This *decision-tree* approach may be satisfactory for small problems, but becomes unwieldy for even medium-sized domains of knowledge. Unfortunately, expert system designers have been less successful in developing general strategies for questioning than strategies for classification.

2.7.2 PRODUCTION SYSTEMS

A general consultation system based purely on the traditional statistical approaches would need the following three components:

1. a data base of sample cases, *the training samples*, which contain the specific statistics for an application
2. a set of general mathematical relationships indicating how the statistics are to be used (e.g. Bayes' rule)
3. the evidence for a new case which is to be classified.

If instead we build a system based on production rules, we have an analogous but more complex situation. Nilsson has described a production system in terms of the following three components:

1. a set of production rules
2. a control system
3. a global data base.

The production rules, or *knowledge base*, are analogous to the training sample statistics. They are the generalized summary of information which can be applied to a specific case. The *control system* (sometimes called the "inference engine") applies the appropriate rules (and stops when a termination condition has been met). The *global data base* is the temporary store of information that is computed for a specific case. This would include the information or evidence directly provided (by the user) about a specific case, and all derived information about the case.

There are many ways in which production systems can be built. The components of the production rules may be relatively elementary items, such as those described in this chapter, or they may be more complicated items such as procedure calls. Production rules are sometimes called *situation-action* rules where the LHS consists of a set of pre-conditions which when satisfied indicate the actions to be taken on the RHS. For most of this book, the actions which we will consider will be represented as classifications and not as procedural knowledge. Although other actions are quite reasonable for many different types of problems, we will concentrate on classifications which are typical of diagnostic problems.

The control system, i.e. the control strategy, must decide when the production rules should be invoked, and it must resolve any conflicts which may occur when several rules are satisfied. There are two principal ways to approach the evaluation of production rules: backward chaining or forward chaining. Backward chaining is often described in terms of goal-directed reasoning or top-down reasoning. Forward chaining is often described in terms of bottom-up or event-driven reasoning.

In backward chaining the system has a set of initial goals, and the rules are invoked in reverse order. The system begins by examining a limited set of production rules, whose right-hand sides are the goals. In a medical domain, for example, they could be the set of final diagnoses. The system then proceeds to examine the left-hand side of the rules to see which of the goals (RHS) are satisfied. As the rules are examined in this backward unraveling, some premises (of the left-hand side of rules) are unknown (logically unsatisfied) and therefore they become new subgoals. If a subgoal is unknown, a question may be asked to determine its status.

In forward chaining the system does not start with any particular goals defined for it. That is, it has no initial subgroup of productions rules which establish a starting point. Instead, the system starts with a subset of evidence and proceeds to invoke the production rules in a forward direction, continuing until no further production rules can be invoked.

Although several systems have been built emphasizing backward chaining, (MYCIN used primarily backward chaining), both forms of invocation and evaluation of the production rules are equally valid as long as they yield the same correct conclusions. The route of arriving at the conclusions will probably differ considerably depending on the strategy adopted. Most classification problems can be solved using either one of the approaches individually or a mixture for production rule evaluation. In Chapter 4 we will describe by example some practical application of these concepts.

Before we look at the practical question of designing an expert system based on production rules, we will briefly examine how several real-world expert systems have been built using variations of the classification model.

2.8 Bibliographical and Historical Remarks

Some of the alternative reasoning methods to be considered in designing an expert system which uses approximate reasoning are discussed in Szolovits and Pauker (1978). A general discussion of the principles of reasoning in an expert system can be found in Buchanan and Duda (1982). Although most expert system designers approach a problem quite differently than the traditional pattern recognition system designers, there is much to be learned from the work in pattern recognition. One of the more readable texts is Duda and Hart (1973). The notion of fuzzy sets was introduced by Zadeh and their use is discussed in Zadeh and Fukanaka (1975). Interest in scoring functions has lessened somewhat since the 1970s. A discussion of the MYCIN scoring function is found in Shortliffe and Buchanan (1975), and the PROSPECTOR scoring function is reviewed in Duda, Hart, and Nilsson (1976). Production rules came into wider use within the AI community following the publication of Newell and Simon (1972). Various AI deduction schemes including production systems are reviewed in Nilsson (1980).

3

Historical Overview of Applications of Expert Systems

3.1 Introduction

In this chapter, we will look at several examples of expert systems applications. Many of the early expert systems were medical consultation systems, and although none of these medical systems found its way into everyday use, the medical systems can provide valuable insight into the various approaches that can be taken in encoding expert knowledge. In several cases, these medical systems have achieved expert performance and have been subjected to more formal evaluations than have other expert systems. Most of these systems used some variant of production rules as the internal representation of the expert's reasoning. However, since production rules are rather elementary descriptive structures, each of these systems has been characterized in other ways than as a pure production rule system. Most of these systems can also be described in terms of a classification model, but some applications have unique characteristics which require extensions to the basic building blocks of production rules and classification models. The generalized systems provide representations which are well understood and which help formalize a problem, at least to the extent of aiding in building a prototype model. Some systems, such as the R1 system (which will be described in section 3.4), are not completely described by a classification model or even by production rules. We will review several prototypical expert systems with special emphasis on those which use the classification model. In selecting examples of expert systems, we restrict ourselves to those which are serious efforts at solving real-world problems, where currently a human expert is required to perform the task.

3.2 Expert Systems in Medicine

In the mid and late 1970s, much of the work in expert systems centered around the development of medical consultation systems. Several consultation systems were developed which exhibited near-expert performance. In contrast to traditional methods of automated diagnosis, these programs did not emphasize performance based on optimal criteria of decision-making such as utility, or information measures. Rather, they attempted to develop representations based on models of the human experts' conceptual structures and reasoning. What follows is a brief review of some of the most highly developed early systems.

3.2.1 CASNET/GLAUCOMA CONSULTATION SYSTEM

During the mid-70s, we developed a modeling approach that used a causal-associational network, CASNET, to represent knowledge of reasoning about diagnosis, prognosis and treatment selection. Although the CASNET system involved a generalized representational scheme and method of reasoning, the main application was in ophthalmology: the diagnosis and treatment of glaucoma, a leading cause of blindness. The mechanisms of glaucoma are sufficiently well understood that they can be used to explain many of the observed patterns of patient findings in terms of causal models. The CASNET model of glaucoma helped provide consultation for complex clinical cases, including those with involved histories and multiple follow-up visits. For some situations it provided alternative opinions derived from different consultants. Using computers at Rutgers University and at Stanford University's SUMEX-AIM facility, a national network of investigators was established to share in the development and testing of the program.

The CASNET or causal-associational network is a particular type of semantic network designed to:

- describe a disease in causal terms
- relate this description to an associational structure of observations (e.g. symptoms and test results)
- describe various classifications imposed on the model (e.g. diagnoses and treatments).

While relations between observations and diagnoses could easily be posed as production rules, the key difference between the CASNET model and the typical classification model lies in the causal model of pathophysiological states. Pathophysiological states are those abnormal conditions typical of a disease process, which, when related as cause and effect, describe the mechanism of the disease. Events are related in the form A *causes* B, and reasoning methods in CASNET take advantage of the causal ordering, although in a somewhat different manner than would be carried out with pure production rules. With causal ordering information, one can deduce the most likely cause that accounts for a particular set of patient data by using an algorithm that traverses the "network" of plausible causes and effects, and finds those causes that, if present, would produce the largest number of observed effects. Unfortunately, in most medical applications there is insufficient knowledge to state events in causal terms. In the CASNET model, observations are used to infer the intermediate causal states. An example would be a rule of the form:

If: The applanation tension is 30 mmHg or higher and
       there is evidence of corneal edema and
       the patient complains of eye pain
Then: *The intraocular pressure must be highly elevated.*

Once the inference of intraocular pressure elevation is made, its causal relationships to other intermediate states are pursued, and when a sufficiently strong causal path has been connected, the diagnostic conclusions are inferred through classification tables or sets of rules that interpret patterns over the causal network. An illustration of the type of reasoning structures used in a CASNET model is given in Figure 3.1.

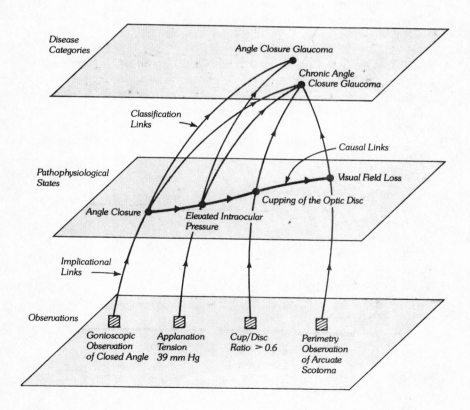

Figure 3.1:   Representation of knowledge in CASNET/Glaucoma

As is typical of most of the medical consultation systems, the user is asked questions and, depending on the responses, additional questions are asked. At the completion of questioning, the system reports its interpretation. A sample session from CASNET/Glaucoma is given next. This session is from the CASNET/Glaucoma system of 1975, so some of the medical advice may no longer be completely up-to-date. As in all examples in this chapter, user responses are printed in boldface and our commentary is in italics.

Example of data entry for right eye (OD) and left eye (OS).

1. Which history topics do you wish to enter:
   (1) Current glaucoma medications (or steroids)
   (2) Symptoms
   (3) Drug intolerances
   (4) Past history of ocular surgery
   (5) Family history of glaucoma
   (6) Patient history of systemic diseases
   (7) Patient history of ocular disease or trauma
   *VALUES: N
2. What is the best corrected visual acuity—20/? (Enter 1, 2, 3, 4 for: FC, HM, LP, NLP respectively)
   *30, 40
3. Enter the intraocular pressure (applanation tension):
   *34, 38
4. Do you have any results from a slit lamp/gonioscopy examination?
   *Y
5. Anterior chamber depth:
   (1) Flat (collapsed)
   (2) Shallow
   (3) Moderate
   (4) Deep
   *2, 2
6. Angles:
   (0) Grade 0—complete or partial closure
   (1) Grade 1—narrow extreme
   (2) Grade 2—narrow moderate
   (3) Grade 3—open
   (4) Grade 4—open
   (5) Slit
   *0, 0
18. Do you have visual field data?
    *Y
19. Visual fields:
    (1) Normal

(2) Nasal step
(3) Temporal step
(4) Isolated paracentral scotoma
(5) Incomplete arcuate scotoma
(6) Complete arcuate scotoma
(7) Arcuate scotoma with breakthrough to periphery
(8) Double arcuate scotoma
(9) Double arcuate scotoma with breakthrough to periphery
(10) Central island
(11) Temporal island
(12) Splitting of fixation

*VALUES:
*OD: 4
*OS: 6

Example of data summary

```
**************************
*  GLAUCOMA SUMMARY  *
**************************
```

Personal data

Name: H.E.
Age: 40   Race: W   Sex: M   Case No: 28

Clinical data summary for visit of 5/5/71

Best corrected visual acuity:
        OD: 20/30   OS: 20/40
IOP:
        OD: 34   OS: 38
Anterior chamber depth:
        Shallow   (OU)
Angles:
        Grade 0   (OU)
Angle closure:
        OD: 150 degrees   OS: 340 degrees
Peripheral anterior synechias:
        OD: 100 degrees   OS: 270 degrees
Horizontal cup/disc ratio: 0.60   (OU)
Optic nerve rim width:
        Variably narrow   (OU)
Visual fields:
        Isolated paracentral scotoma (OD)
        Complete arcuate scotoma   (OS)

```
********************
```

Any revisions (excluding the comments)? N

Example of conclusions from the consultation program

```
*****************************
* DIAGNOSIS AND THERAPY *
*****************************
```

* RIGHT EYE: *

[1]   Present diagnostic status:

Chronic (primary) angle closure glaucoma.

[2]   Treatment recommendations:

Peripheral iridectomy is indicated. Institute weak miotic as needed to normalize tensions. (Add diamox if necessary.)

* LEFT EYE: *

[1]   Present diagnostic status:

Chronic (primary) angle closure glaucoma.

[2]   Treatment recommendations:

Peripheral iridectomy is indicated. Institute weak miotic therapy as needed to normalize tensions. (Add diamox if necessary.)

Sector iridectomy should be considered. Because of a significant degree of PAS, filtering surgery may be considered. However, the increased risk of filtering surgery makes iridectomy preferable. Studies show an equal chance of success of filtering surgery following failure of iridectomy.

Supplementary information for the consultation

Do you want more information (research studies or opinions)?
    *Y

* RIGHT EYE: *

"The medical therapy for angle-closure glaucoma is useful only as a prelude to surgery . . . Miotics and diamox are still most valuable in lowering tension.
The addition of hyperosmotic agents preoperatively should be employed if normal levels of pressure have not been attained."— (Becker-Shaffer, 1970)

"The results of 139 iridectomies performed for narrowness of the angle of the anterior chamber indicate that the operation can be helpful in patients with acute angle closure of recent onset, with prolonged attacks of angle closure, or even with underlying chronic open-angle glaucoma. Intraocular pressure below 25 mmHg with or without medical therapy following iridectomy was achieved in 87% of all cases with acute angle-closure, including 77% cases in which the attack had been present for longer than three days. In 80% of cases with chronic angle-closure glaucoma, and in 80% of cases with chronic open-angle glaucoma that had a superimposed angle closure attack. Subsequent surgery was needed 8% of the cases . . . . The experience gained in the present series of cases indicates that even after synechial closure has developed, iridectomy may still be the procedure of choice in most individuals.'—(Murphy, Spaeth—*Arch. Ophth.*/Vol. 91, Feb. 1974.)

* LEFT EYE: *

"In a recent series of iridencleisis operations on eyes with previous iridectomies there were 9 eyes (26%) that failed to be controlled. This is quite comparable to the average failure rate (26%) of iridencleisis as a primary procedure."—(Becker—Shaffer, 1970) "If lens changes are present, retinal pathology, or a probable need for miotics post-op, then sector iridectomy should be done. If lens changes are severe enough, lens extraction."—(M. Kass, 1976)

The glaucoma consultation system was subject to an intensive evaluation during the 1976 meeting of the American Academy of Ophthalmology and Otolaryngology. The consultation program was used to summarize results of cases and present its recommendations, contrasting them to the opinions of a panel of experts at the Glaucoma Symposium. The panel gave a variety of opinions about the cases, and in almost all of them the program included in its alternatives the main interpretation given by the panel. In a more formal test, the program was made available for testing by the conference attendees. Evaluation questionnaires were filled out by ophthalmologists who were encouraged to test the program with difficult cases. Forty-nine responses were obtained. The tabulation of results showed a 95% acceptance rate for clinical proficiency in the sample questioned. The 77% rate of high competence (the "expert" and "very competent" responses) ascribed to the system by this independent sample of ophthalmologists accorded well with the previously cited judgment of our glaucoma research collaborators.

As with most medical consultation systems, the CASNET/Glaucoma model never became widely used. The problem domain was related in scope to many additional ophthalmological problems which were not covered fully in the original model. Additionally, there was no clear demand for a consultation system in this area. We must recognize that at that time, the mid-70s, most physicians were apprehensive about using computers for consultation. This persists to a somewhat lesser extent today. However, as more and more people, including physicians, are exposed to personal computers in their everyday experience, the fear or distrust of using computer programs that

give advice or consultations will lessen. After our experience and the experience of others in building medical consultations systems, it is clear that the
problem domain must be carefully selected to be one which provides a useful,
and preferably indispensable service to the physician.

Several years of developing the CASNET representation and improving
the CASNET/Glaucoma program to a high level of performance resulted in
the development of new ideas from our own experience as well as comparis-
ons with those of other researchers in expert consultation systems. This led us
to produce a more general representation of knowledge for this type of problem, called EXPERT. While CASNET had proven to be effective for those
problems which could easily be modeled in causal terms, many consultation
problems did not easily fit into a causal framework. Even for those problems
which seem amenable to causal modeling, the effort necessary for describing
the causal model can be quite extensive. The EXPERT system will be discussed in Chapter 4.

3.2.2   MYCIN: INFECTIOUS DISEASE TREATMENT CONSULTANT

The MYCIN system was developed to give advice to a physician about thera-
pies for infectious diseases. Production rules with associated uncertainty
weights served to capture most of the expert knowledge in MYCIN. Production rules are stated in the following typical format:

> **If:**   ($AND (SAME CNTXT INFECT PRIMARY-BACTEREMIA)
> (MEMBF CNTXT SITE STERILESITES)
> (SAME CNTXT PORTAL GI))
> **Then:**   (CONCLUDE CNTXT INDENT BACTEROIDES TALLY .7)

which can be translated as:

> **If:**   The infection is primary-bacteremia and
> the site of the culture is one of sterilesites and
> the suspected portal of entry of the organism is the
> gastro-intestinal tract
> **Then:**   *There is suggestive evidence (.7) that the identity of the organism is*
> *bacteroides.*

The assertions are usually Boolean combinations (e.g. AND) of clauses,
each of which consists of a predicate function (e.g. MEMBF) about a {attri-
bute, object, value} triple (e.g. {SITE, CULTURE-1, STERILESITES}),
where the current context (CNTXT) being considered is CULTURE-1 (as
opposed to CULTURE-2). The triples represent findings and hypotheses
about the patient and related objects or contexts, such as infections, cultures,
or organisms.

The uniformity of representation for expressing domain-specific knowledge
makes it possible for MYCIN to use a general control strategy: a goal-directed

backward chaining of rules. In this approach, the first rule to be evaluated is the one containing the highest level goal: to select treatments for all the infections of the patient. This requires that the infections be known. But since they are usually unknown, the system must then try to satisfy subgoals that will allow the infections to be inferred. Discovering the results of cultures or other clinical parameters of the patient would be the most direct subgoals. These in turn may be deduced from other rules, but eventually the attempt to satisfy rule premises will end with assertions that can only be confirmed by questioning the user for the appropriate information about the patient. Once this happens, the system can begin to reason back by successively satisfying the rules that it had previously left unsatisfied while it was waiting for results. A hierarchical tree of contexts (patient-infections-cultures-organisms) constrains the ordering in which the rules are invoked. These, together with the clinical parameter values and the templates for the parameters, constitute the descriptive component of the MYCIN knowledge base.

Although the MYCIN reasoning strategy is largely based on production rule evaluation, the final selection of therapy is carried out by a detailed algorithm, which uses information deduced by the production rules and other information such as knowledge of which medication covers a specific infection, and knowledge about covering several types of infections with the most efficacious set of medications.

The MYCIN system was formally evaluated by a number of independent consultants. The program performed at a relatively high level, comparable to the experts who participated in the study. An abstracted session of the MYCIN system is given below.

This session is excerpted from Duda and Shortliffe (1983).

The physician wants to determine the most effective combination of antibiotics with which to treat an infection.

In the following initial series of questions, the program decides that the infection is probably meningitis.

Please enter information about the patient.

	Name	Age	Sex	Race
1)	** Pt219	42 YEARS	FEMALE	CAUCASIAN

2) Are there any positive cultures obtained from Pt219?

** NO

3) Are there any pending cultures of Pt219?

** YES

	site	date collected	stain examined?	method of collection [urine, sputum only]
4)	** CSF	3-Jun-75	YES	N/A*

stain of smear   morph
[type NONE if no organism was seen on the smear]

5)   ** NONE

•

•

•

13) Has Pt219 recently had symptoms of persistent headache or other abnormal neurologic symptoms (dizziness, lethargy, etc.)?
** YES

14) Has Pt219 recently had objective evidence of abnormal neurologic signs (nuchal rigidity, coma, seizures, etc.) documented by physician observation or examination?
** YES

The CSF cultures will be considered to be associated with MENINGITIS.

•

•

•

MYCIN has completed its diagnosis. It must now decide on a combination of antibiotics that will cover for the suspected organisms.

My therapy recommendations will be designed to treat for organisms that either are very likely or, although less likely, would have a significant effect on therapy selection if they were presented. It is important to cover for the following probable infection(s) and associated organism(s):

INFECTION-1 is MENINGITIS

    + <ITEM-1>   MYCOBACTERIUM-TB   [from clinical evidence only]
    + ...   COCCIDIOIDES   [from clinical evidence only]
    + ...   CRYPTOCOCCUS   [from clinical evidence only]

43) Does Pt219 have a clinically significant allergic reaction to any antimicrobial agent?
** NO

    [Considering ETHAMBUTAL-AND-INH-AND-RIFAMPIN for use against INFECTION-1...]

44) Is Pt219 pregnant or breast feeding?
** NO

45) Do you have reason to suspect that Pt219 may have impaired renal function?
** NO

46) Pt219's weight in kilograms (or <number> POUNDS) :
** 51.3

[REC-1] My preferred therapy recommendation is as follows:
Give the following in combination:
1) ETHAMBUTAL
Dose: 1.28g (13.0 100mg-tablets) q24h PO for 60 days [calcu-
lated on basis of 25 mg/kg] then 770 mg (7.5 100mg-
tablets) q24h PO [calculated on basis of 15 mg/kg]
[The dose should be modified in renal failure.] I can adjust
the dose only if a creatinine clearance or a serum creatinine
is obtained.
Comments: Periodic vision screening tests are recommended for
optic neuritis.

2) INH
Dose: 513 mg (5.0 100mg-tablets) q24h PO
[calculated on basis of 10 mg/kg]

3) RIFAMPIN
Dose: 600 mg PO q24h
Comments: Administer dose on an empty stomach

Although I suspect that fungus may be a possible pathogen, in the absence of definitive evidence, I will not recommend antifungal therapy at this time. Please obtain samples for fungal, TB, and viral cultures, cytology, VDRL (blood and CSF), coccidioides complement-fixation (blood and CSF), cryptococcal antigen (blood and CSF), viral titers (blood and CSF). An infectious disease consult may be advisable.

The MYCIN system has been generalized to EMYCIN (*Essential MYCIN*) which is application independent; a reasoning model can be developed for many problems that are suited to the production rule representation of EMY-CIN. EMYCIN has been used to build consultation programs in several domains: pulmonary diseases, structural analysis, and computer fault diagnosis.

3.2.3 INTERNIST-I: DIAGNOSTIC CONSULTANT IN INTERNAL MEDICINE

The INTERNIST-I system (its successor is known as CADUCEUS) is a large system being developed for diagnosis in internal medicine. Because the number of possibilities is very large, knowledge is encoded in a highly structured and relatively straightforward representation. The INTERNIST-I system has been reported to cover a large proportion of the field of internal medicine, and has been tested with many complex cases reported in the major medical journals.

Production rules are not used; instead INTERNIST-I represents most of its medical knowledge in weighted links between findings and diagnoses. The *evokes* weight from finding F to diagnosis D approximates the (subjective) inverse probability, P(D|F), that D is present given that F is observed, and used for example to express the probability that a person has had a heart attack given severe chest pain. The *frequency* link from diagnosis D to finding F approximates the (subjective) direct probability, P(F|D), that F will be observed when D is present, and used, for example, to express the probability that a person who has a heart attack will have severe chest pain. The system also contains a taxonomic tree of diagnosis categories in which terminal nodes represent specific diseases which the patient can have and non-terminal nodes represent more general disease categories such as liver disease. Associated with each finding is a global *import* weight which is a subjective measure of how important it is to explain or account for the finding if it is present. Findings are grouped into categories such as signs and lab tests to indicate the degree of difficulty associated with obtaining their values. Generally, more easily obtained findings are requested first during a diagnostic session.

The model builder using INTERNIST-I generates rules primarily by establishing relations between pairs of findings and diagnoses and estimating the weights to be used by the system. Modification of an established model (e.g. to correct its performance on an incorrectly diagnosed case) would then consist of adding new links between existing F-D pairs previously unrelated, adding new findings or diagnoses, or adjusting the weights associated with existing links between findings and diagnoses.

INTERNIST-I's diagnosis weighting strategy can be characterized as a scoring function that computes the weight that the patient has a disease, given the set of present findings, by adding and subtracting weights associated with findings. For each disease being considered, the system adds the evokes (F-D) weight of each present finding which is evidence for that disease, subtracts the import of findings that are present and unexplained, and subtracts the frequency (D-F) weight for findings that are absent but are expected for that disease. A bonus is added for each previously confirmed diagnosis causally linked to the current diagnosis. The questioning strategy used during a consultation session changes according to the size of the set of likely diagnoses and the relative diagnosis weights within the set. During one stage of diagnosis the system tries to find new evidence for diagnoses that have already been suggested by asking about new findings that would suggest that diagnosis if they were found to be present. If the system accumulates enough likely disease candidates it changes its questioning strategy to distinguish among several competing diagnoses by asking about findings that would distinguish one diagnosis from another. If one diagnosis has a much higher weight than others the system may change its questioning strategy again and pursue it by asking about the expected findings if that diagnosis were present, i.e. findings linked to it via D-F links. If the difference between the most highly weighted diagnosis and the next highly weighted diagnosis surpasses a threshold, the highest weighted diagnosis is concluded. In all questioning modes, however, the weighting strategy remains additive. INTERNIST-I does not use explicit rules which deny diagnoses in its knowledge base. INTERNIST-I's successor system, CADUCEUS, is still an active project, and extensions are being made to its representation to handle denial of hypotheses and production-like conjunctive rules. INTERNIST-I attempts to use this representation not only for classification of a single disease, but also for differential diagnosis and classification of multiple problems. It uses a strategy of question selection which tries to cover all potential findings. The relatively simple representation of INTERNIST-I has led to quite protracted question sessions in trying to cover multiple possibilities. A major effort of current CADUCEUS program development is to extend the representation to allow for more directed questioning.

An abstracted sample session with the INTERNIST-I system is described next.

This session is excerpted from Miller, Pople, and Myers (1982).

INTERNIST-I consultation   SUMEX-AIM Version
15-May-81 07:31:39
ENTER CASE NAME: NEJM-CASE-30-1969-ADMISSION-1

The user enters the initial positive findings and negative
findings from a list of several thousand observations.

INITIAL POSITIVE MANIFESTATIONS:
+AGE GTR THAN 55
+ARTHRITIS HX
+DEPRESSION HX
+SEX FEMALE
+THYROIDECTOMY HX
+ULCER PEPTIC HX
+URINE DARK HX
+WEIGHT INCREASE RECENT HX
+ANOREXIA
+CHEST PAIN LATERAL EXACERBATION WITH
  BREATHING
+CHEST PAIN LATERAL SHARP
                              •
                              •
                              •
+DENY

INITIAL NEGATIVE FINDINGS:
— ALCOHOLISM CHRONIC HX

•

•

•

— T4 TOTAL BLOOD INCREASED
— GO

GO indicates the end of the presentation of initial findings and allows the system to begin questioning.

INTERNIST-I indicates that it is temporarily disregarding some findings which it cannot yet explain and is concentrating on two diseases which can explain many of the findings.

DISREGARDING: DEPRESSION HX, WEIGHT INCREASED RECENT HX . . .

CONSIDERING: AGE GTR THAN 55, SEX FEMALE, URINE DARK HX, ANOREXIA, FECES LIGHT COLORED, JAUNDICE, LIVER ENLARGED MODERATE . . .

DISCRIMINATE: HEPATITIS CHRONIC ACTIVE, BILIARY CIRRHOSIS PRIMARY

The system is focusing its attention on discriminating among these conditions.

Please Enter Findings of LIPID(S) BLOOD
GO   *ask individual items of this question*
CHOLESTEROL BLOOD DECREASED?
N/A   *not available*
Please Enter Findings of KIDNEY FUNCTION TEST(S)
GO
SKIN XANTHOMATA?
NO
Please Enter Findings of BLOOD COUNT AND SMEAR
GO
RBC TARGET(S)
NO
Please Enter Findings of KIDNEY FUNCTION TEST(S)
GO
UREA NITROGEN BLOOD 60 to 100?
NO
UREA NITROGEN BLOOD GTR THAN 100?
NO
Please Enter Findings of LIVER FUNCTION TEST(S)
GO

UROBILINOGEN URINE ABSENT?
N/A

CONCLUDE: HEPATITIS CHRONIC ACTIVE

The system has reached a diagnosis. Several findings, however, are not explained by the diagnosis, and the system turns it attention to these remaining findings.

DISREGARDING: DEPRESSION HX, CHEST PAIN LATERAL
EXACERBATION WITH BREATHING, . . .

CONSIDERING: WEIGHT INCREASE RECENT HX,
ABDOMEN DISTENTION, ABDOMEN FLUID WAVE, ASCITIC
FLUID OBTAINED BY PARACENTESIS, ASCITIC FLUID
WBC 100 TO 500, ASCITIC FLUID PROTEIN 3 GRAM(S)
PERCENT OR LESS

CONCLUDE: TRANSUDATIVE ASCITES

•
•
•

3.3   PROSPECTOR: An Expert System in Geology

PROSPECTOR is an expert system designed to aid in mineral exploration. Communication with the system can be described as mixed initiative: either the system requests information from the user, or information is volunteered by the user using key words. The system will then attempt to indicate the likelihood of various mineral deposits. Models of several different ore deposits have been developed including a sulfide model, a carbonate lead/zinc model, a porphyry copper model, a nickel sulfide model, and a uranium model.

The model is primarily one of production rules with additional structures to encode the various semantic and taxonomic relationships. A simple example of a rule in the system is the following:

If: There are abundant quartz sulfide veinlets with
         no apparent alteration halos
Then: *There is alteration favorable for the potassic zone.*

Confidence measures are assigned to the production rules and a scoring function is used to combine the weights. At any given time, PROSPECTOR focuses its questions on the inference rule that affects its current goal the most, which is analogous to pursuing the most likely hypothesis.

Components of the production rules, i.e. the findings and hypotheses, are represented as relations between entities and their properties. Examples of relations are AGE, COMPOSITION, FORM, LOCATION, SIZE and TEXTURE. Examples of the relation values include specific rock formations, physical locations, times and sizes. Examples of the entities are GEOLOGIC-AGES, ROCKS and MINERALS. Taxonomic relationships between various entities are also described such as: PYRITE is a SULFIDE.

In a typical session, the user will begin by offering a set of field observations. The inference strategies are somewhat similar to those of MYCIN. As the production rules are invoked, findings and hypotheses are either deduced, or else a question is asked. Each piece of evidence is examined to determine the one which, if it were know, would affect the odds on the hypothesis the most.

In order to constrain the order in which questions are asked, a context may be described for a piece of evidence. If A is a context for B, then A must be established prior to B.

A model design system, KAS (Knowledge Acquisition System), has been developed to aid in the construction of new models for PROSPECTOR, in the mineral deposit domain and other domains. Most of the models developed for PROSPECTOR require modification and refinement over a long period of time. Because the inference rules and the semantic and taxonomic links are closely related, a change to one will affect the other. KAS was designed to aid the model designer as much as possible in knowledge acquisition.

Since much information must be entered for each new rule or hypothesis, KAS helps the user keep track of needed information. For example, the system will ask the user to supply any rules or hypotheses which have been referenced but not defined. It will also request input of omitted confidence measures or other descriptions, and will perform syntax checks on the entered data. KAS will also supply defaults where possible, taking into account the network structure. An abstracted session is given below. In this example for the nickel model, the user responds to most questions by typing a number from 0 to 5, with 5 indicating a high degree of confidence. Other responses are a simple yes or no, or one of the allowable keywords. Some models such as the uranium model rely less on a question selection strategy. Instead, the user fills out a questionnaire which contains a list of all useful information. The object of the analysis is to compute a favorability function on a scale of 0 to 5, indicating the likelihood of the presence of the mineral which is modeled. This favorability index corresponds to the internal confidence measures accumulated by the PROSPECTOR scoring function in the course of invoking the model's production rules. The PROSPECTOR group has also developed models that process map data. In such cases the data for each zone is analyzed separately by the model and eventually a favorability map is constructed. The map is shaded to indicate the degree of favorability of finding the indicated mineral. The PROSPECTOR geological system is not in everyday use. However, in one instance, the system was able to identify a mineral site of potential commercial significance which had been overlooked by a mining company. A session with PROSPECTOR system is described below.

This is an abstracted session with the KNSD, Komatiitic-nickel-sulfide model (Duda, Hart, Barrett, Gaschnig, Konolige, Reboh, and Slocum 1978). For some questions the user responds with a degree of belief measured on a scale of 0 to 5, with 5 indicating strong belief.

1—To what degree do you believe that:
(The target area lies in a greenstone belt)? **5**

2—To what degree do you believe that:
(The target area is in a region containing significant volume of ultramafic rocks)? **5**

3—What is the age of the greenstone belt? **ARCHEAN**
What is your confidence in this estimate? **4**

4—Can you identify more than one discrete volcanic cycle within the greenstone belt? **NO**

5—To what degree do you believe that:
(The prospective ore body lies in a sequence of mafic to ultramafic igneous rocks)? **4**

6—To what degree do you believe that:
(This sequence of mafic to ultramafic rocks forms a Komatiitic rock suite)? **?**

This KNSD model uses the extended definition of Komatiite rock suite as proposed by Arndt et al. (1977) rather than the other more restrictive definitions formulated by Viljoen (1969) and Brooks and Hart (1974). In case you are uncertain as to whether you are dealing with a Komatiite rock suite, many of the following questions are designed to clarify this point. To what degree do you believe that the sequence of mafic to ultramafic igneous rocks forms such a Komatiite rock suite? **0**

7—To what degree do you believe that:
(Some olivine-rich rocks are present as part of this sequence of mafic to ultramafic igneous rocks)? **0**

8—To what degree do you believe that:
(This sequence of mafic to ultramafic igneous rocks contains noncumulate textured peridotite)? **0**

9—To what degree do you believe that:
(This sequence of mafic to ultramafic igneous rocks contains noncumulate textured peridotite)? **0**

10—To what degree do you believe that:
(This sequence of mafic to ultramafic rocks contains pyroxene rich rocks)? **5**

.
.
.

At the end of the session the system displays the favorability index for a nickel deposit. The result was less than 1, relatively unfavorable.

3.4 R1: An Expert System for Configuring Computers

R1 (sometimes referred to as XCON) is an expert system which configures VAX systems for Digital Equipment Corporation. Its input is a customer's order and its output is a set of diagrams depicting the spatial relationships between the components, together with indications of the components which are missing but necessary to complete a system. A typical VAX system contains dozens of major components, such as a CPU, memory control units, and an assorted collection of cabinets, peripheral devices, drivers for the devices, and cables. Only certain components can be attached to one another, and this limits the possible combinations that might be considered. Knowledge about how to attach these components is represented by several thousand production rules. R1 includes in its knowledge base a list of potential properties (attribute-value pairs) of each of the hundreds of components that could go into building a VAX system. These will vary according to each VAX order.

Problem solving in R1 primarily involves the matching of production rule arguments. As with most production rule models, control is achieved by having certain items in the data memory called goals. When a production attempts to achieve a goal it generates the necessary subgoals to be processed. Unlike the previous systems we have examined, a questioning strategy is not required. Rather, the initial list of components is given and then R1 invokes the production rules and generates the VAX configuration. After a production rule is satisfied, the action required by the specific production rule is taken and the production rules are re-evaluated to see what further action is necessary. As the production rules are satisfied, the configuration is gradually built until a satisfactory configuration is found, or it is discovered that components are missing. No confidence measures are used by R1. However, the system has a scheme for conflict resolution, because several rules may be satisfied simultaneously. An example of a an R1 production rule is given in Figure 3.2.

> **If:**   The current context is assigning devices to unibus and
> there is an unassigned dual port disc drive and
> the type of controller it requires is known and
> there are two such controllers
> neither of which has any devices assigned to it and
> the number of devices that these controllers can support
> is known
>
> **Then:** *Assign the disk drive to each of the controllers*
> *and note that the two controllers have been associated*
> *and that each supports one device.*

Figure 3.2:   Example of R1 production rule

A typical R1 goal is:

(CONTEXT-ACTIVE G00087 SELECT-RIGHT-SIZE-BACKPLANE 316)

This goal requires the system to select a backplane to put in the box G00087 on the basis of the amount of space remaining in the box and the unibus modules remaining to be configured. Goals are satisfied by having productions that generate the necessary subgoals when a match occurs. The subgoals are processed before the higher level goal.

The production rule interpreter, programmed in the OPS language, has access to two memories: production memory and data memory. Production memory is a place to store the productions and any linear ordering between productions. Data memory is a place to store the data processed by the productions and any static relations between the data. The interpreter of the production rules repeatedly matches the productions against a subset of the information held in the data memory, selects one or more productions with satisfied antecedent conditions, then executes the selected production and effects changes to the production and data memories. This sequence of operations is called the Recognize-Act Cycle. This cycle basically consists of Match, Conflict Resolution, and Action components.

In the Match component of the cycle, every legal instantiation of every production is found. In conflict resolution, a single production is allowed to execute from the set of legal instantiations found during Match. The OPS interpreter has specified priorities to determine which of the match instantiations should be used if more than one is found. For example, the instantiation of productions with the greatest number of condition elements are given preference. In the Action stage, once a production is chosen, it is executed by individually executing each of its sub-actions (subgoals) in order. R1 is an example of a system in everyday use. Every VAX order is processed through R1; thousands of orders have been examined by R1. Of particular value is the determination by R1 that a component may not have been included in the order. An abstracted example of an R1 session is given next.

This computer output is excerpted from McDermott (1980).
The customer order is presented.

COMPONENTS ORDERED:

1 SV-AXHHA-LA	[packaged system]
1 FP780-AA	[floating point accelerator]
1 DW780-AA	[unibus adaptor]
1 BA11-KE	[unibus expansion cabinet box]
6 MS780-DC	[memory]
1 MS780-CA	[memory controller]
1 H9602-HA	[cpu expansion cabinet]

1 H7111-A	[clock battery backup]
1 H7112-A	[memory battery backup]
1 REP05-AA	[single port disk drive]
4 RP05-BA	[dual port disk drive]
1 TEE16-AE	[tape drive with formatter]
2 TE16-AE	[tape drive]
8 RK07-EA	[single port disk drive]
1 DR11-B	[direct memory access interface]
1 LP11-CA	[line printer]
1 DZ11-F	[multiplexer with panel]
1 DZ11-B	[multiplexer]
2 LA-36-CE	[hard copy terminal]

This is the result of R1's analysis.

SUBSTITUTIONS
NONE

COMPONENTS ADDED
1 H7101
1 M9014
1 M9202
1 M9302
1 070-11528

THE FOLLOWING COMPONENTS WERE NOT CONFIGURED

2 RK07-EA*	(POSSIBLY-FORGOTTEN-PREREQUISITE DISK-DRIVE CONTROLLER)
1 H7111-A	(NOT-NEEDED)
1 H7100-A	(NOT-NEEDED)
6 BC06R-10	(NOT-NEEDED)
2 070-12292108	(NOT-NEEDED)
1 BC05F-15	(NOT-NEEDED)
2 BC11A-10	(NOT-NEEDED)

POSSIBLY FORGOTTEN COMPONENTS
3 BC05L-15 TO CONNECT UBA 0 TO BOX 1 OF CABINET 1
1 RK611* TO SUPPORT 2 UNUSED DISK-DRIVE

UNUSED CAPACITY
THE MEMORY CONTROLLER (MS780-CA*) IN THE CPUX CABINET COULD SUPPORT 768 K BYTES MORE MEMORY
THE MEMORY CONTROLLER (MS780-CA*) IN THE CPU CABINET COULD SUPPORT 768 K BYTES MORE MEMORY

DZ11-D* 1 IN BOX 2 OF CABINET 1 COULD SUPPORT 7 MORE
MA20 LINES (THROUGH PANEL 2)
DZ11-D* 2 IN BOX 2 OF CABINET 1 COULD SUPPORT 7 MORE
MA20 LINES (THROUGH PANEL 2)
DZ11-B* 2 IN BOX 1 OF CABINET 1 IS UNUSED (IT CAN SUP-
PORT 8 EIA LINES (THROUGH PANEL 1))
DZ11-B* 1 IN BOX 1 OF CABINET 1 IS UNUSED (IT CAN SUP-
PORT 8 EIA LINES (THROUGH PANEL 1))
MBA 0 COULD SUPPORT 2 MORE MB-DEVICES AND ITS 1 MAS-
TER TAPE DRIVES COULD SUPPORT 5 MORE SLAVES
MBA 1 COULD SUPPORT 4 MORE MB-DEVICES

•
•
•

*R1 next displays a series of drawings which indicate how the various
machine components may be assembled and configured in the working
computer system.*

3.5 DART and DASD: Expert Systems in Computer Fault Diagnosis

In recent years a number of computer manufacturers have been examining
the use of expert systems for improving diagnosis of computer system failures.
Several approaches have been explored, but one which appears promising in
the near-term fits well within the classification model. This approach has
much in common with medical diagnosis, and the experience gained in the
fault diagnosis systems should prove useful for many different types of complex equipment repair. DART is the overall project name for a number of
computer fault diagnosis projects at IBM in collaboration with Stanford
University. A major long-term research effort of the DART project is to capture a deeper form of knowledge by including a detailed functional model of
a computer system. We will discuss a system which restricts itself to production rules. One of the initial prototype systems reported in the literature has
been a system designed for fault diagnosis in the teleprocessing units of IBM
computers, which we will refer to as TP. This model was originally implemented in the EMYCIN representation. The object of these fault diagnosis
models is to find the cause of a specific malfunction at the major component
level. This is a relatively high level decision, as opposed to fault diagnosis at
the circuit level.

The TP model uses a pure production rule scheme. Questions are asked
and, based on the responses reported to the system, related production rules
are invoked. This process continues until all related questions are asked. The

conclusions of the system are then displayed indicating the most likely cause of the failure. Some of the information that is asked could actually be retrieved directly from a dump or error log, or a data base indicating configuration characteristics and a past history of problems. More recently, after the original work on the teleprocessing unit, developmental efforts have concentrated on the DASD system for fault diagnosis of disk devices. Our research group has assisted IBM in implementing a version of the DASD model using the EXPERT formalism. The DASD model has performed well in informal field tests at one site. However, it is a promising prototype, not an expert system in routine use.

Many of the production rules require that some specific information be retrieved and processed before further productions rules are invoked or further questions asked. For example, if it is known that the device is an IBM 3380 storage device, then the production may need bits 0−7 of a particular register, whereas if the device is a 3370, different bits and registers may be required. The DASD model differs somewhat from the pure classification models, such as those found in medical diagnosis, in that many of the production rules actually require some intermediate action, such as examining certain bit patterns and recording those patterns which are important for the diagnosis. In medical diagnosis most information is in the form of direct observations or intermediate conclusions, but these are gathered by the user and not by the system. A production rule for a medical problem usually takes no action other than referencing or reaching a conclusion, whereas the DASD model requires that actions be taken by the system, other than just producing a classification. This is illustrated, in Figure 3.3, by an abstracted rule from

If: The storage device is a 3380 and
         a state save or dynamic trace has been performed
Then: *The pack change byte = bits 8 through 15 of*
         location INBR of the trace.

If: The class of device that is failing is 3380 and
         the problem is missing interrupt and
         a dynamic trace is available and
         the trace indicates major hardware, microcode, or
         formatting errors and . . .
         The failing interface has a pack change interrupt
         outstanding . . .
Then: *The HOST is the cause of the failure with a confidence of .9.*
         The missing interrupt is caused by the device being reserved
         without being released.

Figure 3.3:   DASD production rules

the DASD model. An example of a session in the EMYCIN version of the TP model program for diagnosing faults in the teleprocessing unit of IBM computers is given below.

This session is excerpted from Bennett and Hollander (1981).

1) What is your user-number?
** **909776**
2) Are you familiar with the use of this tool?
** **YES**
3) Briefly describe the reason for this session:
** **HAVING DIFFICULTY ACTIVATING TP NETWORK**

•
•
•

The first system for SESSION-201 will be called:

----------SYSTEM-1----------
5) Is SYSTEM-1 a loosely compiled multi-processor?
** **NO**
The first problem with SYSTEM-1 will be called:

----------PROBLEM-1----------
Please enter the following information.

CPU machine type	SCP name	SCP running under VM?
6) ** 3081	MVS/SP	NO

7) What system is exhibited by PROBLEM-1?
** ?
    Expected responses are: UNABLE.TO.LOGON.
    UNABLE.TO.ACTIVATE.TP.NETWORK.MISSING.DATA.
    HANG.LOOP.etc.
** **UNABLE.TO.ACTIVATE.TP.NETWORK**
8) Is the NETWORK.ACTIVATION problem occurring
    in a SNA network? (YES NO)
** **YES**
9) Is the terminal(s) involved in PROBLEM-1
    locally (channel) attached, remotely
    attached or both? (LOCAL REMOTE BOTH)
** **REMOTE**

The system begins to ask questions about the TP subsystem.

•

•

•

Complete one line for each remote terminal.

		applic.	communic.
terminal	cluster	program	controller
LU-name	PU-name	LU-name	address

LU=logical unit, PU=physical unit

13) ** **TERM1   CLUS01   IMS   059**
14) ** **DONE**
The path to terminal TERM1 will be called:
----------SNA.PATH-1----------
15) What is the name of the LINE connecting
    CLUS01 to 059 (its TP controller)?
** **LINE01**

•

•

•

18) Please enter the name of the access
    method used by IMS to communicate with
    TERM1. (ACF/VTAM VTAM ACF/TCAM TCAM)
** **ACF/VTAM**
19) What type of TP controller is used
    for this terminal? (3704 3705)
** **3705**
20) What control program is running in this
    TP controller? (NCP PEP)
** **NCP**

•

•

•

22) Please indicate ALL of the nodes which
    are able to achieve and maintain
    (throughout the activation process) an
    active status.
** **IMS**

To most effectively identify the cause of this
    problem will require a trace showing the
    activation of the LINE, the 3705 and NCP.

23) Which VTAM trace types are available to
aid in pursuing this problem?
** **VTAM.IO**
24) Does your VTAM IO (RNIO) trace show
that an ACTLINK (activate link)
command was sent to line LINE01?
** **YES**
25) What does your VTAM IO (RNIO) trace
indicate was the response to the
ACTLINK? (POSITIVE NEGATIVE NONE)
** **NEGATIVE**

.
.
.

Eventually the system concludes that a line problem is causing the failure

The most likely source(s) of the failure discussed in PROBLEM-1 is:

LINE-LINE01   (.5)

> The VTAM IO (RNIO) trace indicates that an ACTLINK was sent
> to NCP for LINE01 and that a negative response was received. The
> ACTLINK should have caused NCP to issue an ENABLE to the line
> and this, in turn, should have caused the modem to respond with
> DATA SET READY (DSR). You might want to check the modem
> interface. If the DSR is active then it is likely that the line is not the
> source of the failure. A LINE (or PT2) trace could be used to further
> investigate the problem.

.
.
.

3.6   Putting Expert Systems in Perspective

The definition of an expert system given in Chapter 1 is not meant to be a
rigorous and complete definition. Rather, we have described certain types of
expert systems whose representations may be generalizable to other domains.
The expert systems in medicine which have been used as examples do fit the
mold of classification problems and do have easily generalizable structures
which are well understood. Several of these systems have undergone formal
evaluation. While most of these programs are not in everyday use, the results
of the evaluations do support the contention of expert-level performance in
circumscribed areas. The use of the medical systems is restricted by many
non-technical factors such as legal and social issues. There are other artificial
intelligence systems which have achieved relatively high performance. Many

people would describe these systems as expert systems. Two in particular come to mind: MACSYMA, which is a system for symbolic integration, and DENDRAL, a system for inferring the molecular structure of unknown chemical compounds using mass spectrometry data. These systems are quite complex, and do perform well, but they are not easily representable as generalizable systems which give clear practical directions for someone who is building a system in an entirely different domain. They appear as highly developed, specialized systems, the forerunners of other systems which have led to the generalized practical ideas that have emerged in the past few years. Not everyone will agree that an expert system should be implemented in a generalized framework. The simplest criterion for expert system performance is that the system achieve a high level of performance which matches, but does not necessarily exceed, the human expert's performance. Very often the program that has been implemented using a generalized expert system tool can be recoded as a conventional program. However, there are many advantages to using a generalized tool, most important of which is the ease of maintaining and modifying the knowledge base.

If we examine the group of systems chosen as examples in this chapter, we see some trends and patterns in the types of systems that have evolved. These can give us direction when we wish to develop practical systems for new applications. For example, all the medical consulation problems are classification problems. While none of the medical systems described is in routine use and only INTERNIST-I (CADUCEUS) is still under active development, these systems are representative of the various approaches researchers have taken, and these medical systems have achieved verifiable expert performance. The INTERNIST-I system is clearly of very broad scope, but relies for its reasoning on a straightforward additive scoring method that uses many, relatively shallow reasoning rules. It has some similarities to probabilistic schemes, but does not attempt to be optimal mathematically. The possibility of relying on many direct reasoning rules makes sense in an extremely broad application, such as all of internal medicine, but further, detailed analysis would likely require much more complex reasoning and specialized expertise.

The CASNET system was, in some ways, a system that was ahead of its time. Causal models are now undergoing a revival in academic research circles as an area for intensive research. The CASNET system represented a practical approach to building a deeper reasoning model that would underlie or support an associational, production rule model for consultation. A relatively high performance and elegant model was built using this approach. However, while this model uses causal reasoning very effectively, it is not clear that such reasoning is more satisfactory to the end user, or that the physician routinely uses causal reasoning for diagnosis or treatment. The model designer must determine whether one should invest the additional effort necessary to acquire the causal knowledge and to represent this knowledge in the computer. The cost/benefit analysis may not be favorable yet for the model designer who is building a practical system. Nor is the design process for causal models as well understood as that for the much simpler production rule models.

MYCIN was an early example of a production rule model and more recent models such as R1 continue to use the production rule approach. While MYCIN and PROSPECTOR are classification models, R1 is not a pure classification system. We may, however, try to view R1 as a variant of the classification model where intermediate results are highly changeable, being more of a dynamic classification model. In R1, intermediate decisions are made, i.e. actions are taken, and the model is re-evaluated to see what the next decision should be. In the typical classification model, the situation is much more stable: new information is gathered, questions are answered, but few direct actions are taken by the system, other than synthesizing information supplied by the user.

In the remainder of this book, we will emphasize the classification model and production rule approach. Those *generalizable* expert systems which have achieved success in limited domains have relied on production rules. While these systems employ relatively shallow reasoning, they do seem to work well for many diagnostic and interpretive problems.

Many of the important research problems of the 1970s for expert systems are still unresolved. Yet, while many difficulties remain, we now have the ability to offer solutions to some of these issues when we consider applications in well-circumscribed domains. Our perspective on what is important and necessary in designing a specialized expert system has changed dramatically since the early days. Many researchers, however, would maintain that the same research issues remain and that only limited progress has been made in resolving those issues that are fundamental to the building of expert systems. Before we begin to address some of the technical issues of building a practical system, we should first describe our views on some of the methodological issues of expert system development.

Most of the artificial intelligence community uses LISP for developing their systems. Many people feel that LISP's capabilities for symbolic computation make this the language of choice. A program written in LISP, however, is not by definition an expert system. The user of an expert system is interested in what the system does, not the programming language in which it was written. Although LISP is particularly powerful for symbolic processing and several powerful LISP-based programming environments have been developed by the AI community, other languages may be considered for practical goals such as programming familiarity and efficiency. These languages may require additional programming effort, but it quite possible to implement an expert system in a language other than LISP. Our work has been done mostly in FORTRAN, and this has made communication with the world outside the university much easier than it might otherwise have been. Skilled knowledge engineers and programmers, who understand the key ideas of designing expert systems, can choose from many higher level languages. One recent approach is to use PROLOG, a logic programming language.

The general tools for building expert systems can do a good job on certain types of problems. For example, EXPERT is particularly adept at solving classification problems. One still needs a skilled individual to design the model and represent it in the system. Not all problems will fit the system. Does that mean that such a system fails on the problem? Not necessarily. It may mean that some additional code may have to be written and added to the general development system to take care of the special needs of certains problems not envisioned by the designers of the system. One must bear in mind that these systems are developmental tools. Their most important purpose is to provide an orderly means of representing and using knowledge in a prompt manner. These systems are tools for formalizing expert knowledge. They provide a means of building a prototype and experimenting with this knowledge. Once a model is built and performs well, or even before then, it may be possible for a more efficient, but less flexible scheme to be used, i.e. a special purpose expert system may be programmed. By using one of the generalized tools, with its accompanying discipline of model-building, one is given direction when one would otherwise be in the dark at the beginning of a project. Farther along in the project one may see the possibilities and develop an improved approach, more tailored to the application. The generalized systems may be viewed as tools for experimentation and prototype building. However, for some problems they may be quite satisfactory as they stand, and in other cases, certain small changes in their code may be sufficient to produce an acceptable system.

The skills necessary for building an expert system are not identical to those for general programming. Building an expert system usually requires excellent computer programming skills, but additional skills are also needed. Unlike many programming projects, an expert system does not usually lend itself to detailed design specifications. Rather, the system is constantly changing and is often subject to redesign. The changes may be major, particularly during the early stages of development of the system. The designer of such a system must be able to adapt and innovate. Surely someone using a highly developed general system for designing an expert system may require fewer programming skills; but using a generalized system tool is not an ordinary programming task. The skills of designing an expert system are much more artistic than purely mechanical. Instead of programming a solution to a problem, we are often taking a problem that is defined in one way by the expert initially and quite differently later as he gains experience with the expert system prototypes. Besides the actual programming, the more important issue is that of knowledge acquisition, i.e. getting knowledge from the expert and encoding that knowledge in the computer. The knowledge engineer must become quite knowledgeable about the area of application, an area in which he may have no prior experience.

Many researchers believe that efficient knowledge aquisition from the expert is the bottleneck in building expert systems. To a large extent this is true, but knowledge acquisition is not an issue that can be easily separated from all other issues of design of expert systems. We would like to have the expert work alone building the expert system himself, but with the current state-of-the-art this is not feasible. Building an expert system can be interesting, but difficult work, requiring many long hours of interviewing the expert. The best we can do today to ease the task of knowledge acquisition is to use some general framework that allows us to rapidly build a prototype model, and get feedback from the expert on its performance. Because of the effort required to build a state-of-the-art system, it is important that the initial selection of the human expert be a good one. Not all experts are amenable to providing the structured knowledge needed by an expert system. The knowledge acquisition task is even more difficult, if one does not have the ability to show some initial results, which then provide the basis for more pointed comments by the expert.

The time frame for developing an expert system clearly depends on the application. The Serum Protein Diagnostic Program took less than a year. This relatively brief period was due to circumstances which were especially advantageous: the right combination of problem, model designers, experts, programmers, and past experience. An expert system is a living project, one that should grow and improve in scope and performance as time passes. There should be landmarks in the performance of the system. Many problems will prove much more complex than the serum protein problem. However, after a year, one should at least see a running prototype that performs relatively well in a circumscribed domain. This prototype should be more than a running system for a single hardwired example: it ought to cover at least a representative spectrum of typical, though possibly simplified, problem cases.

None of the systems we have described has a real natural language capability. There are numerous applications where very little input is necessary from the user. The Serum Protein Diagnostic Program and R1 are examples. While one may argue that natural language understanding is important for many tasks, natural language is not very precise, and an expert system requires specific information for its reasoning. A natural language understanding capability would add another layer of uncertainty to the performance of the expert system, since the system could have inaccurately interpreted the information reported by the user. If we consider natural language processing to be essential for some expert systems, it is still quite possible to view the task of building the expert system knowledge base and reasoning procedures as somewhat independent. The natural language component may be interfaced at some later time.

Expert systems have no general mechanisms for *common sense* reasoning, and are generally appropriate for only highly circumscribed domains. However, expert systems may still be extremely valuable. Many expert systems are designed to be intelligent assistants, and it is an intelligent user of the system who provides the common sense. The system can often provide information about interpretations that might be given by various experts for a particular situation. The system might provide alternatives to a user who wants to double check his results. But the user must recognize that he is still in control, that the system may make mistakes, and that he may dismiss the conclusions of the system. For example, a system which give advice on a computer failure can be extremely valuable to many field engineers. They may use the program because they are having a difficult time finding the cause of the failure, or they might just want to review the alternatives.

In the mid-70s there was a great deal of interest in devising heuristic scoring functions that would be somewhat analogous to probabilities, while not suffering from their non-human mathematical constraints. It was believed that such mechanisms would be at the heart of any high performance expert system. Several schemes were analyzed carefully, such as the MYCIN and PROSPECTOR scoring functions. These scoring functions are heuristic approximations which are not probabilistically exact, but are related to certain other constraints or assumptions about reasoning. If the assumptions are not met, then major inaccuracies can occur. However, knowledge for many applications can be represented without using uncertainty measures. This was the case with the R1 project. The Serum Protein Diagnostic Program required only a very limited set of confidence measures. Often, the confidence measures are used as a shorthand to rate the relative importance of one production rule versus another, rather than a means of providing accurate probabilities or measures of uncertainty.

The first priority of the system is that it must reach correct conclusions. A model that reaches expert conclusions without explanation may still be a valuable tool. However, a form of explanation is built into most expert systems that use production rules. The system usually can easily cite those rules which were successfully satisfied, and therefore led to a particular conclusion. Clearly, this is valuable in debugging a model and examining why an incorrect conclusion was reached. Whether the same explanation is satisfactory to the ultimate user as opposed to the system developers is a subject of much discussion in the research community. Learning from experience has proven difficult for the current generation of systems. Much of the practical work in expert systems relies on the knowledge engineer to capture expertise and encode it on the computer. If the program makes a mistake once, the same mistake will be made again, unless the program is changed by the knowledge engineer. Learning from experience is an important research issue in artificial intelligence, but so far most approaches have proven impractical for large-scale models. As is the case with automating knowledge acquisition, learning from experience is an important research issue that will enhance the performance of expert systems. Yet, current systems which do not have these capabilities can still prove valuable.

Let's turn our attention from reviewing existing systems to a specific illustration of how one might go about designing an expert system.

3.7 Bibliographical and Historical Remarks

An excellent collection of articles on the early medical expert systems has been edited by Szolovits (1982) and reviews can be found in Shortliffe, Buchanan, and Feigenbaum (1979) and Kulikowski (1980). The CASNET/Glaucoma program is described in Weiss, Kulikowski, and Safir (1978), and the generalized CASNET approach for building causal models is described in Weiss, Kulikowski, Amarel, and Safir (1978). The MYCIN approach is detailed in Shortliffe (1976), and the results of an evaluation of the program's performance are reported in Yu et al. (1979). The INTERNIST-I system and its evaluation are described in Miller, Pople, and Myers (1982). The PROSPECTOR system has been described in Duda, Gaschnig, and Hart (1979) and PROSPECTOR's generalized knowledge acquisition system is described in Reboh (1981). R1 is discussed in McDermott (1982), and the OPS programming language in Forgy and McDermott (1977). In addition to OPS, several other generalized approaches to building nonclassification expert systems have been attempted including AGE (Nii and Aiello 1979) and Hearsay-III (Balzer, Erman, London, and Williams 1980). The early work on the DART project was reported by Bennett and Hollander 1981). Recent research on the DART project, emphasizing detailed functional models of computer systems, has been presented in Genesereth (1982). The MACSYMA system is described in MACSYMA Mathlab Group (1977). The DENDRAL programs have been used extensively by chemists for many years and numerous articles have appeared both in the computer science and chemistry literature. A summary of its application can be found in Buchanan and Feigenbaum (1978).

4

Designing an Expert System

4.1 Introduction

The classification model has proven to be an excellent representation for expert system problems of diagnosis or interpretation. Several general expert system tools have been developed using the classification model. This approach is relatively well understood and classification systems will have many similarities in design. In many instances, it may be possible to program the expert system as a conventional program. Yet, because the task of designing a specific application "model" usually involves extensive experimentation and modifications, it is likely that a more general approach is necessary to the design of the system. Once the logic of the application model is worked out, the model can be fine tuned to the specialized needs of the application. The general systems for developing expert models can be viewed as experimental tools for rapidly building a running system. At a later stage, the developed system may then be converted to the specialized format needed to run routinely.

While a classification system may employ productions rules, the form of the production system can be tailored to classification problems. The classification model, illustrated in Figure 4.1, may have the following three components:

- conclusions
- observations
- production rules.

In our discussion, we will also use the term "model" to refer to a specific application of the classification model.

A production system, illustrated in Figure 4.2 consists of these three components:

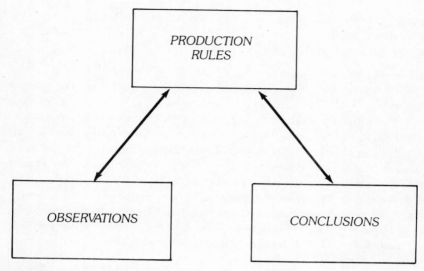

Figure 4.1:   Classification model

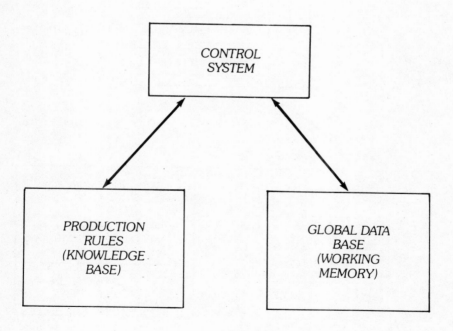

Figure 4.2:   Production system

- a set of production rules
- a control system
- a global data base.

These concepts were introduced in Chapter 2. However, if we build a production system for classification problems, we may have a system as illustrated in Figure 4.3. The knowledge base may be organized as a classification model. For a given case, the working memory will consist of recorded values of observations and degrees of belief in various conclusions. The control system will determine how and in what order the production rules are to be evaluated. The control system will try to resolve conflicts, and may request that further information should be provided.

There are many variations of systems which may fit the mold of Figure 4.3. We will not try to cover many of these variations. Instead we will choose an example from a real expert system tool, the EXPERT system which we have developed. EXPERT is a relatively simple general system which has been

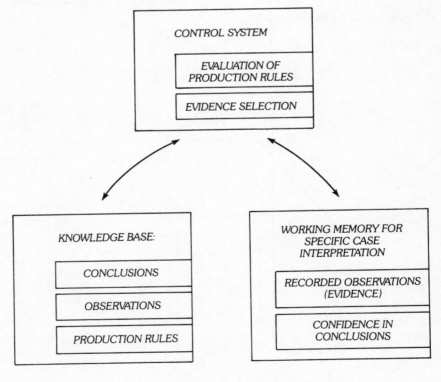

Figure 4.3: Classification model combined with production system

used successfully in numerous applications including medicine, oil well-log analysis, laboratory instrument interpretation, and computer fault diagnosis. Even within this system there are many alternatives which we will not cover. Our purpose is not to give a detailed view of a single expert system, but rather to cover some of the broader practical issues of building a classification expert system.

EXPERT is one among several expert system design tools. It is one of the simpler systems, and does illustrate that if a system is specialized for a class of problems, such as classification problems, one can achieve some very important design goals including:

- Ease of model design. Prototype models should be capable of running in relatively brief periods of time.
- Efficient system performance.
- Predictable performance. It should be relatively easy to understand the interaction among decision rules.
- Empirical testing. The performance of a model should be matched to a data base of stored problem cases providing for verification and consistency-checking of a model.

In the following examples, we describe a simple hypothetical problem in automobile maintenance and repair. A model of a car that can't start is presented as an illustration of a prototypical expert systems type problem. With this example, we will try to cover many of the practical issues of designing an expert system. This example will not cover these issues in a detailed manner, and the discussion is not meant to be a primer on the EXPERT language. The examples, however, are taken directly from a running model written in the EXPERT representation. The examples are hypothetical in the sense that the model is not truly expert in the domain, although it is likely that a model could be developed along these lines to the point of expert performance. First, let's look at what might be a typical session using this model. Note that upper and lower case characters may be interchanged in the syntax of the language.

4.2 A Sample Consultation Session

The computer will ask questions of the user to gather information about the problem. In this case questions are posed using a straightforward format, involving multiple choice questions, checklists, or individual numerical or truth-value (yes/no/unknown) responses.

Enter Name or ID Number: Cadillac
Enter Date of Visit: 5/1/82

1. Type of Problem:
     1) Car Won't Start
     2) Other Car Problems
   Checklist:
   *1

2. Simple Checks:
     1) Headlights Are Dim
     2) Fuel Filter Clogged
     3) Battery Cables Loose or Corroded
   Checklist:
   *2

3. Starter Data:
     1) No Cranking
     2) Slow Cranking
     3) Normal Cranking
     4) Grinding Noise From Starter
   Checklist:
   *Why

The user wants to know why this question was asked. This question includes several related items about the starter which are always asked together. The program indicates which item it is most interested in obtaining:

If: There is a grinding noise from starter
Then: *Conclude starter malfunction (.9).*

3. Starter Data:
     1) No Cranking
     2) Slow Cranking
     3) Normal Cranking
     4) Grinding Noise From Starter
   Checklist:
   *Fix 2

An incorrect entry was made for question 2

FIX: Simple Checks:
     1) Headlights Are Dim
     2) Fuel Filter Clogged
     3) Battery Cables Loose or Corroded
     Checklist:
     *1,2

3. Starter Data:
     1) No Cranking
     2) Slow Cranking

    3) Normal Cranking
    4) Grinding Noise From Starter
Checklist:
*1

4. Gas Gauge Reads EMPTY
   *No

5. Odor of Gasoline in Car:
    1) None
    2) Normal
    3) Very Strong
   Choose one:
   *2

The reported results are summarized:

SUMMARY

Name: Cadillac
Case: 5   Visit: 1   Date: 05/01/82

Type of Problem:
   Car Won't Start

Odor of Gasoline in Car:
   Normal

Simple Checks:
   Headlights Are Dim
   Fuel Filter Clogged

Starter Data:
   No Cranking

The conclusions are printed with associated measures of confidence. These numbers are not a probabilities but analogous to them: the closer the number is to 1, the greater the confidence.

INTERPRETIVE ANALYSIS

Diagnostic Status
   1.00   Fuel Filter Clogged
   0.90   Battery Discharged

Treatment Recommendations
   0.80   Replace Gas Filter
   0.80   Charge or Replace Battery

The user wants to know why the conclusion "battery discharged" was reached. DX2 indicates the appropriate mnemonics which can be referenced by the HYPO command.

:DX2
Diagnostic Status

    FILT   1.00 Fuel Filter Clogged
    BATD   0.90 Battery Discharged

Treatment Recommendations

    RFILT   0.80 Replace Gas Filter
    GBATT   0.80 Charge or Replace Battery

:hypo(batd)
Battery Discharged was set by rule 7 in this manner:

If: Starter Data: No Cranking
         Simple Checks: Headlights Are Dim
Then: Battery Discharged (0.9).

By retrieving the appropriate rule that led to the conclusion, we or the expert can understand the model's reasoning, and change it if necessary by modifying the model.

4.3   Describing the Expert Knowledge

How was the above consultation produced? What form of EXPERT model was needed? In the EXPERT language, three representational components are used to design a consultation model:

- hypotheses or conclusions
- findings or observations (data)
- reasoning or decision rules.

Unlike EMYCIN or PROSPECTOR, in EXPERT a sharp distinction is made between findings and hypotheses. Findings are observations or measurements. These findings are reported in the form of true, false, numerical, or unavailable responses to questions. Hypotheses are the conclusions that may be inferred by the system. A measure of uncertainty is usually associated with a hypothesis. The reasoning or decision rules are expressed as production rules, or If-Then statements.

The alternative approach used by other systems, such as EMYCIN or PROSPECTOR is to describe hypotheses or findings as triples of object, attribute, value. An example of such a triple would be "the color of the car is green." In this example the object is the car, the attribute (or logically speaking the predicate) is color, and the value is green. While these triples are somewhat richer in structure than the simple findings and hypotheses that we will use in our examples, either approach may often be used for classification systems. In logical terms, EXPERT works mostly at the simpler propositional logic level, whereas EMYCIN and PROSPECTOR include many expressions of the predicate (functional) logic level.

In the following sections, examples from the auto repair model will illustrate each of these structures.

4.3.1 EXPRESSING THE CONCLUSIONS THE EXPERT MIGHT REACH

Let's first look at the hypotheses, or conclusions that may be inferred by the system, and which therefore set the bounds on the area of expertise to be covered. In a medical application these might be diagnoses or therapy recommendations. In many other cases, these conclusions can represent any form of advice or interpretation. Depending on which observations and measurements, are made, different degrees of uncertainty may be associated with a hypothesis. In EXPERT, each hypothesis is described by a short-hand mnemonic, and a full statement in natural language (English or whatever other language the designer wishes to use). The mnemonics are used to reference the hypotheses when writing the decision rules. In the simplest form, hypotheses are just presented as a list, though in more complex models we can specify hierarchical relations between them. For example, what follows is a list of simple car problems which we want to express in the model.

FLOOD	Car Flooded
CHOKE	Choke Stuck
EMPTY	No Fuel
FILT	Fuel Filter Clogged
CAB	Battery Cables Loose or Corroded
BATD	Battery Discharged
STRTR	Starter Malfunction

A major goal in the design of a model is to abstract the reasoning of the expert. It is important to reason not only with hypotheses that represent the final conclusions of the expert, but also with intermediate hypotheses or conclusions. These typically summarize results from many, related measurements, or may simply be a qualitative abstraction of an important piece of evidence. Once defined, they can be used to make the reasoning clearer and more efficient. It is much easier to reason with a smaller set of intermediate hypotheses than a much larger set of all possible combinations of findings. For example, because there may be many types of fuel system problems, an intermediate hypothesis FUEL could be created to summarize whether a problem of this type has occurred. This hypothesis may then be referenced in reasoning rules. In our model, there are four fuel system problems: *car flooded, choke stuck, no fuel, and fuel filter clogged.*

Example:

The *taxonomy declaration describes the conclusions the system may reach. Although taxonomic relations could be described, we will examine the simplest structure: a list of conclusions.

**hypotheses
*taxonomy

FLOOD	Car Flooded
CHOKE	Choke Stuck
EMPTY	No Fuel
FILT	Fuel Filter Clogged

*intermediate hypotheses

FUEL	Fuel System Problems
ELEC	Electrical System Problems

Rules could be now written to conclude the intermediate hypothesis FUEL, a fuel system problem, on the basis of information about FLOOD, CHOKE, EMPTY, or FILT.

Additional hypotheses could be used to describe the types of advice that may be given about actions to be taken. For example, these hypotheses could be added to the model.

Example:

*TREATMENTS

WAIT	Wait 10 minutes or Depress Accelerator to Floor while Starting
OPEN	Remove Air Cleaner Assembly and Manually Open Choke with Pencil
GAS	Put More Gasoline into Tank

RFILT	Replace Gas Filter
CLEAN	Clean and Tighten Battery Cables
GBATT	Charge or Replace Battery
NSTAR	Replace Starter

4.3.2 Expressing the Observations Needed To Reach Conclusions

Findings are the observations or measurement results needed to reach conclusions. They are usually described as having a logical or "truth" value such as: true, false, or unknown; or as having a numerical value. The mode of interaction usually consists of the system requesting results from the user. However, there are cases where the finding can be recorded without direct user intervention, such as a direct reading from an instrument or a result sent from another program, which is then recorded as a finding of the expert system. For the logic of the expert system, it is sufficient that the finding be supplied from some source. There are therefore many ways in which findings can be reported: sequentially, in batch mode, by interactive questioning, or from user-volunteered information. If the model is to ask questions in an organized fashion, some structure for organizing the questions must be adopted. Certain designers advocate restricted forms of natural (English) language interaction. This can be technically complicated, prone to errors, and somewhat imprecise. The expert system works best with precise findings, which may not be what it is given in natural language. However, the reporting of findings is clearly a place where a natural language interpreter may be interface to an expert system.

Even without considering a particular control strategy, such as data-driven or goal-driven control, we can structure the findings by related topics, so that questioning can proceed more effectively. An efficient means of communication, and one which has proven highly effective, has been to structure questions into menu-like groupings. This scheme allows the questions to be organized by topic as multiple choice questions, checklists, or numerical response questions. Checklists are sets of questions for which any number of responses (including none at all) within the set may be valid. The simple yes/no response type of question can also be effective in controlling the sequence of questioning when used to control question sequencing, although its misuse can lead to asking too many questions. These simple question structures are often quite reasonable for organizing the question topics, yet require a minimum effort on the part of the user in communicating the results. An example of how we might structure the questions for the car example is as follows:

*multiple choice
Odor of Gas in Carburetor:

NGAS   None
MGAS   Normal
LGAS   Very Strong

*Checklist
Type of Problem:
FCWS   Car Won't Start
FOTH   Other Car Problems

*Numerical
TEMP   Outdoor Temperature (degrees F):

*yes-no
EGAS   Gas Gauge Reads Empty

The findings as presented are sufficiently well-defined to use in the decision rules. No indication is given about how hard it is to obtain the information for the finding. Some systems treat findings as hypotheses and every finding may have a level of certainty associated with the finding. For example, the user could indicate that he is 90% sure that the temperature is 55° or 70% sure that the odor of gas in the carburetor is normal. In our example system we will limit responses to be true, false, unknown, or a numerical value. However, with this scheme, the model designer may decide that the degree of certainty for a specific finding is important, by creating a new finding to measure factors beyond the actual result. For example, one may request an instrument reading and then also ask about the reliability of the reading.

Although findings can express most of the information needed for the antecedents of reasoning rules that work directly from evidence, in some cases the model designer may find that the model also has to include more detailed procedural knowledge. In effect, the model must call a subprogram to perform a task which will return with a finding. For example, one could have a finding representing miles per gallon of gasoline. This could be posed as a question to the user. On the other hand, it may prove more natural to have a procedure compute the miles per gallon from the actual mileage and fuel usage. Conceptually, one can usually proceed with the design of a reasoning model with the understanding that some findings will have to be obtained by means other than the interactive questioning of the user of the system. During the prototype design phase, however, it is often simplest to enter the data through a *questionnaire*.

A list of questions does not imply any particular ordering in which they need be asked. That is left to a strategy of question selection which will be discussed later. However, there is often additional information which can simplify strategy considerations considerably. The designer of a model may have a very good idea of how the questions should be asked, either because of the conventions of his specialty, or through experience. In these cases a large part of the question ordering can be specified in advance. It is possible to group several of the question types (e.g. checklist, numerical) together in the form of a questionnaire. When any member of the questionnaire is selected to be asked, *all* the questions must be asked in the order of the questionnaire.

*Begin questionnaire

*Checklist
Car Lights not working:

FRONT	Front of Car
REAR	Rear of Car

*Checklist
Front Lights not working:

HEAD	Headlights
FTURN	Turn Signals
PARK	Parking Lights

*Checklist
Rear Lights not working:

TAIL	Tail Lights
RTURN	Turn Signals
BU	Backup Lights

*End Questionnaire

Figure 4.4:   Example of a questionnaire

If the question strategy wants to determine any of the car light findings of the questionnaire of Figure 4.4, the system must ask all of these questions in the given order. For example, if the system wants to know whether the headlights are not working, it must begin by asking "car lights not working" and proceed in order to the other questions. The advantage of this type of question structure is that we can organize information by topic and maintain a coherent dialog with little computing overhead on the system. Perhaps a deficiency of this approach is that the system may ask additional questions which are not immediately important for the current case.

In the next section, we will see how the questionnaire can be used in conjunction with very simple production rules which selectively branch within a questionnaire. All questions in the questionnaire may not be asked if the system is able to determine that the answer to some questions can be directly deduced from previous findings.

4.3.3 EXPRESSING THE REASONING RULES

Production rules are by far the most common form of representation for the decision rules. These are the If-Then rules used for compiling the *rule-of-thumb* reasoning procedures of the expert. In terms of our representation which includes hypotheses and findings, the production rules can be categorized in terms of the three types of logical relationships between findings and hypotheses:

- FF — finding-to-finding rules
- FH — finding-to-hypothesis rules
- HH — hypothesis-to-hypothesis rules.

4.3.3.1 FINDING-TO-FINDING RULES

FF (finding-to-finding) rules specify truth values of findings that can be directly deduced from an already established finding. Because more powerful forms of production rules can be described combining both findings and hypotheses, the use of FF rules is usually limited to establishing local control over the sequencing of questions. The FF rules specify deterministic logical relations between findings whose truth values have already been determined, and other findings whose truth values have not yet been obtained. In this way we can avoid asking unnecessary questions if the answers have already been determined by the response to a previous question. These rules are most advantageously used within a questionnaire consisting of questions which proceed from the general to the specific. As in Figure 4.4, we can compose questionnaires which group questions so that they are asked in strictly consecutive order, from first to last. We can then specify conditional branchings which, at any given stage, depend on previous responses to sections of the questionnaire. The following example illustrates this idea.

If: The *finding* front lights not working is *false*

Then: *All findings regarding front headlights not working are false, i.e. HEAD, FTURN and PARK.*

This can be stated in the EXPERT syntax as follows:

*FF rules

F(FRONT, F) → F(HEAD:PARK, F)

Note that "F(FRONT, F)" means finding FRONT is false. "F(HEAD:PARK, F)" indicates all findings from HEAD to PARK inclusive in the order given in the listing are false.

Similarly, a rule can be composed for the rear lights:

F(REAR, F) → F(TAIL:BU, F)

In this example, only if we are told that the front lights are not working will the questions about specific front lights be asked. Similarly, only if we are told that the rear lights are not working will the questions about specific rear lights be asked. This example shows how to design a questionnaire that will guide the sequence of questioning in a natural fashion—limiting its inquiries for data to those that haven't been ruled out by some previous response.

4.3.3.2 FINDING-TO-HYPOTHESIS RULES

Many of the classification model systems have been designed with production rules that allow for a measure of belief in a conclusion. A belief measure is typically a number that ranges from -1 to $+1$, and has fewer mathematical constraints associated with it than a probability measure. The value of -1 would stand for complete lack of belief or confidence in the conclusion, whereas a $+1$ stands for complete confidence. A zero stands for an undecided or unknown level of confidence. A major difference between most confidence measures and probabilities lies in the separation between statements of certainty (belief) and uncertainty (disbelief) in a hypothesis. Probabilities require that the probability of a hypothesis is always one minus the probability of its negation. Confidence measures allow the expert to assign beliefs and disbeliefs on an informal basis which is not tied to a frequency interpretation for each of the reasoning rules. It does not, however, exclude such an interpretation of the overall results of the model.

An advantage of using any of the reasonable confidence measures comes from their ability to express the expert knowledge more concisely than would be the case if confidence measures were not used. There are applications, however, which can be solved quite well without confidence measures or only with full confirmation and denial of hypotheses.

It is sometimes useful to distinguish FH (finding to hypothesis) rules from the more powerful HH (hypothesis to hypothesis) rules, because a rule consisting of only findings can be handled more simply and efficiently. Figure 4.5 shows how we can simply combine a pair of findings to infer a hypothesis. Figure 4.6 illustrates a shorthand notation for combining a finding with a set of alternatives. By writing [n:A,B,C,D], we mean that *n or more* of the alternative findings A,B,C,D must be satisfied for the entire condition to be satisfied, or evaluated as "true". In Figure 4.6, where $n=1$, this allows us to state the fact the any one of the list is sufficient. Without such a notation we would have to write many separate rules.

*FH Rules

F(SCRNK,T) & F(DIM,T) → H(BATD,.7)

If: The starter cranks slowly and the headlights are dim
Then: *The battery is discharged (with a confidence of .7).*

Figure 4.5:   Elementary finding-to-hypothesis rule

F(TEMP,0:50) & [1:F(SCRNK,T),F(OCRNK,T)] → H(CHOKE,.6)

If: The temperature is between 0 and 50 degrees
         and 1 or more of the following is true:
         starter cranks slowly or doesn't crank at all
Then: *The choke is stuck (with a confidence of .7).*

Figure 4.6:   FH rule using disjunction

4.3.3.3   HYPOTHESIS-TO-HYPOTHESIS RULES

The HH (hypothesis to hypothesis) rules allow the model builder to specify inferences between hypotheses. In contrast to EMYCIN and PROSPECTOR, in EXPERT a hypothesis specified in an HH production rule is stated for a fixed interval of confidence. The advantages and disadvantages of this approach are described in section 4.4.1. The following is a simple example of an HH rule.

*HH Rules

F(FCWS,T) & H(FLOOD,.2:1) → H(WAIT,.9)

If: The car won't start and
         we have concluded that the car is flooded
         (with a confidence between .2 and 1)
Then: *Wait 10 minutes or Depress Accelerator to Floor while Starting
         (confidence of .9).*

While the car repair example is a very small prototypical system, it is often the case that an expert system may consist of hundreds or even thousands of rules. Practical considerations such as efficiency, modularity, and ease of description suggest an additional descriptive component for the production rules: *context* descriptors. The concept of context is to limit the invocation of a group of production rules to specific situations. Only when preconditions have been satisfied will a group of production rules be considered. In the EXPERT syntax, a group of HH rules is partitioned into two parts. The *If* conditions must be satisfied before the *Then* condition rules are considered. For the following example, only if the finding FCWS is true, i.e. car won't start, will the grouped production rules be evaluated.

```
*HH Rules
*If
F(FCWS,T)
```

Note: This is shorthand to factor out the common expression F(FCSW,T). This condition also determines a "context" for evaluating the next list of rules. If FCWS is false, the remaining rules will not be evaluated.

```
*Then
H(FLOOD,.2:1) → H(WAIT,.9)
H(CHOKE,.2:1) → H(OPEN,.5)
H(EMPTY,.3:1) → H(GAS,.9)
H(FILT,.4:*) → H(RFILT,.8)
H(CAB,.5:*) → H(CLEAN,.7)
H(BATD,.4:1) → H(GBATT,.8)
H(STRTR,.4:1) → H(NSTAR,.9)
*END
```

We have completed our description of a model that has many of the features of the kinds of expert models which can be described using production rules. The examples are taken for a running prototype model using the EXPERT representation. The following is the completed prototype model from which the sample session was produced.

```
**Hypotheses
*Taxonomy
FLOOD    Car Flooded
CHOKE    Choke Stuck
EMPTY    No Fuel
FILT     Fuel filter clogged
CAB      Battery Cables Loose or Corroded
BATD     Battery Discharged
STRTR    Starter Malfunction

*Intermediate Hypotheses
FUEL     Fuel System Problems
ELEC     Electrical System Problems

*Treatments
```

WAIT   Wait 10 minutes or depress accelerator to floor +
            while starting
OPEN   Remove air cleaner assembly and manually open choke +
            with pencil
GAS   Put more gasoline into tank
RFILT   Replace gas filter
CLEAN   Clean and tighten battery cables
GBATT   Charge or replace battery
NSTAR   Replace starter

**Findings

*checklist
Type of problem:
FCWS   Car Won't Start
FOTH   Other Car Problems

*multiple choice
Odor of gasoline in carburetor:
NGAS   None
MGAS   Normal
LGAS   Very Strong

*checklist
Simple checks:
DIM   Headlights Are dim
CFILT   Fuel Filter Clogged
LCAB   Battery Cables Loose or Corroded

*checklist
Starter data:
NCRNK   No Cranking
SCRNK   Slow Cranking
OCRNK   Normal Cranking
GRIND   Grinding Noise From Starter

*numerical
TEMP   Outdoor Temperature (degrees F):

*yes/no
EGAS   Gas Gauge Reads Empty

**Rules
*FH Rules

*Note: When 2 or more rules that are not mutually exclusive conclude the
same hypothesis, a mechanism is needed to combine the weights or resolve
conflicts in the weights. This will be discussed in section 4.4.1.*

F(LGAS,T) → H(FLOOD,.8)
F(TEMP,0:50) & [1:F(SCRNK,T),F(OCRNK,T)] → H(CHOKE,.6)
F(EGAS,T) → H(EMPTY,.9)
F(CFILT,T) → H(FILT,1.0)
F(LCAB,T) → H(CAB,.9)
F(SCRNK,T) & F(DIM,T) → H(BATD,.7)
F(NCRNK,T) & F(DIM,T) → H(BATD,.9)
F(NCRNK,T) & F(DIM,F) → H(STRTR,.7)
F(GRIND,T) → H(STRTR,.9)

*HH Rules
*If
F(FCWS,T)

*Note: This is shorthand to factor out the common expression F(FCSW,T). This
condition also determines a "context" for evaluating the next list of rules. If
FCWS is false, the remaining rules will not be evaluated.*

*Then
H(FLOOD,.2:1) → H(WAIT,.9)
H(CHOKE,.2:1) → H(OPEN,.5)
H(EMPTY,.3:1) → H(GAS,.9)
H(FILT,.4:*) → H(RFILT,.8)
H(CAB,.5:*) → H(CLEAN,.7)
H(BATD,.4:1) → H(GBATT,.8)
H(STRTR,.4:1) → H(NSTAR,.9)
*END

4.4   Using the Knowledge in a Consultation Session

The representation described in section 4.3 is sufficient to represent much of
the core reasoning structure of an expert system. The prototype auto repair
model is quite typical of classification models and the sample session is
directly produced from this model. In a prototypical application, there are
two major issues of control in the design of an expert system. These are the
related goals of:

- reaching accurate conclusions
- asking reasonable questions which aid in the interpretation.

Building an expert system is far from a precise science. The expert often
provides a voluminous amount of information; we must try to extract the key
components of his reasoning and try to be as precise and concise as possible in

representing this knowledge. Because there are many variations among existing implementations of production rules, it is very important to choose those structures and strategies which suffice for the application at hand. For example, there are many ways of expressing strategies of questioning. However, for a given application the questioning order may not be important or the strategy of questioning may easily be specified in advance for any specific case. This was illustrated with the use of questionnaires for the auto repair model. Within the questionnaire, control is handled by very simple mechanisms such as FF rules. As we shall see, problems which require reasoning under uncertainty are more complex, and the addition of confidence measures will produce a somewhat more complicated analysis.

4.4.1   RANKING AND SELECTION OF CONCLUSIONS

First let's look at the situation where we have a set of production rules which have been specified by the model designer. The user of this model will enter a case, as was illustrated in the sample session of the auto model. It is likely that only a small subset of the production rules specified in the model will be satisfied by any given case. In it's simplest form, the ranking of the conclusions is merely a list of all the hypotheses that were satisfied, perhaps ordered from the highest to the lowest weights above a specified threshold. In the sample session for the auto repair model, we saw the following conclusions:

INTERPRETIVE ANALYSIS

Diagnostic Status

    1.00   Fuel Filter Clogged
    0.90   Battery Discharged

Treatment Recommendations

    0.80   Replace Gas Filter
    0.80   Charge or Replace Battery

The hypothesis *Battery Discharged* was concluded, because for this case the following production rule was satisfied:

    If: Starter Data: No Cranking
              Headlights Are Dim
    Then: *Battery Discharged (0.9).*

This form of analysis describes the general approach to rule evaluation. There are, however, several complicating factors. We must be concerned with the effect of the order of evaluation of the rules. In almost all realistic problems the order in which the evidence is gathered should not alter the conclusions. So it follows that the same conclusion must be reached from a given set of findings regardless of the order in which rules are evaluated. If all production rules were like the FH rules, their order of invocation would indeed never alter the conclusions. This holds because FH rules don't interact, and the left hand side of the rule consists solely of findings which at any given moment may be true or not. Rules like the HH rules, however, which are typical of most production rule systems, often depend on intermediate results which come from the invocation of other rules. For example, the following rule was described for the auto repair model:

F(FCWS,T) & H(FLOOD,.2:1) → H(WAIT,.9)

>   **If:** The car won't start and
>   we have concluded that the car is flooded
>   (with a confidence between .2 and 1)
>   **Then:** *Wait 10 minutes or Depress Accelerator to Floor while Starting.*

The hypothesis that *the car is flooded* must be established prior to the invocation of this rule. There are several ways of dealing with this problem. In EXPERT the rules are ordered by the model designer, so that the order in which the HH rules are listed in the model is the actual order in which the rules are evaluated. Each rule is evaluated only once for each cycle of reasoning in the consultation. A reasoning cycle begins when new information (findings) is received and all the HH rules are re-evaluated. This is a relatively straightforward scheme that lends itself easily to implementation and presents no inherent ambiguities. A potential drawback of this approach is that the expert must provide an ordering of the rules. In our experience this has not been a serious problem. An alternative rule evaluation approach is to iterate through an unordered set of rules until a steady state is achieved, i.e. no changes in confidence occur for a complete pass of evaluation of the rules.

One of the more popular alternatives to the ordered rule approach is known as *backward chaining* of hypotheses. This approach is less efficient and somewhat harder to implement, but it does not require the pre-ordering of rules in a model. When a rule is evaluated using backward chaining, the rules for each hypothesis appearing in the left hand side of the rule are examined to see whether the hypothesis is satisfied. The chain of rules is tracked down as far as necessary to determine whether the hypotheses in question are satisfied. The backward chaining approach assumes that no loops occur among the rules; dependencies between hypotheses appearing in the same rule may result in ambiguities of results. For example, a rule of the form

$$F(A,T) \& H(B,.1:1) \rightarrow H(B,.9)$$

would not be valid, because there is a loop in the logic. This type of rule, however, may make sense to the model designer, and would be valid if the rules were invoked using the ordered scheme.

A major purpose of the confidence measures is to enable the model designer to express the uncertainty with which the expert expresses his knowledge. Perhaps not immediately apparent but equally important is another use of confidence measures or weights: the abstraction of the expert's knowledge into a concise representation. There are often huge numbers of production rules needed to cover the entire logical spectrum of possible combinations of findings and hypotheses for even a small model. For instance, a model with only 15 yes/no findings and 2 hypotheses would potentially involve more than 2000 rules. We therefore would like to have an approach which inherently reduces the number of production rules which must be specified to capture the expert's knowledge. The confidence measures and a scoring function which adds up the weights for a given case can be useful aids in abstracting the expert's knowledge. If we express all possible combinations of findings and hypotheses in a mutually exclusive manner, there would be no problem. Only one production rule could be satisfied for a given case. Even for small models, this is an unreasonable and unnecessary approach. We are usually interested in only a small subset of production rules which are critical for reaching the appropriate conclusions. However, it would be difficult and unnatural to express these production rules as being mutually exclusive. We therefore must expect that several rules may be satisfied at any stage in the cycle of reasoning, and that these rules may imply the same hypothesis with varying confidence measures, some of which may appear to be in conflict. Here is an example from the auto repair model where different rules may be satisfied and these rules are applicable to the same hypothesis.

Example:

F(NCRNK,T) & F(DIM,F) → H(STRTR,.7)

  If: The starter doesn't crank and
           the headlights are not dim
  Then: *There is a starter malfunction (with a confidence of .7).*

F(GRIND,T) → H(STRTR,.9)

  If: There is a grinding noise from the starter
  Then: *There is a starter malfunction (with a confidence of .9.)*

If both rules are satisfied in this example, we must be able to combine the .7 and .9 measures into a single number. In this case both numbers are in the positive direction. However, confidence measures may range from from -1 to 1 and this must be taken into account for any scoring function which is used to combine the confidence measures.

The scoring function which we have used extensively and which has worked well in our experience is one of the simplest. We choose the the maximum absolute value of confidence in that hypothesis. Thus, the confidence assigned to the conclusion *starter malfunction* would be .9 when both the .7 and .9 rules were satisfied. (For negative values, the absolute value is used in the comparison, e.g. -1 wins over .9.) The concept of a more complicated scoring function that somehow combines the weights of several rules is very attractive. Even in this case, one might argue that we should combine .7 and .9 into a weight greater than .9. We should, however, consider many factors other than our intuition about the validity of a scoring funtion. These production rules, in general, are not mathematically complete. They are rules of thumb and do not satisfy certain critical logico-mathematical constraints such as mutual exclusivity, or normalization of the measure (on a scale of 0 to 1) over the logical alternatives. Whatever scoring function is used will be a heuristic that may work in some cases, but in others will yield poor and inaccurate results. The problems of scoring functions are related to the practical difficulties of applying mathematical and statistical methods to large-scale inference problems. These difficulties are exacerbated in the case of production rules, because the model is so different from one used according to rigorous mathematical formulations.

The scoring function's purpose is more than that of just approximating probabilities. We would instead like to write rules which allow us to come up with reasonable confidence weights and preserve the relative ranking of conclusions. The conclusion that is considered more likely will be ranked higher than other, less likely conclusions, even if the numbers are not exact probabilities. There is no magic to a scoring function, and most of the scoring functions that use reasonable assumptions yield results as good as any other. There are dangers, however, in scoring functions, particularly the more complicated ones. The major problem is that with several rules being satisfied it is very difficult for the model designer to track down how the final weight for a conclusion is reached. An even more difficult problem, once the rules involved in reaching the final weight for a conclusion are known, is the task of trying to manipulate the numbers to arrive at the correct ranking of conclusions. With a complex model, this is like juggling hundreds of balls at once. That is why we like to use as overall confidence measure the one with the maximum (absolute) value. It usually points in the right direction of inference. But more importantly, with it we can more easily understand how the weights were assigned to the conclusions. It is relatively easy to figure out how things can go wrong, and to make corrections to the rules when necessary.

We will look briefly at alternatives to the maximum (absolute) confidence for use as a scoring function. One alternative scoring function is that used in the MYCIN system. In its scheme, when a new rule is satisfied for a hypothesis with confidence CF, the new weight for the hypothesis is updated as follows:

New Weight = Current Weight + (1 − Current Weight) * CF

In the previous example for the *starter malfunction*, when the first rule was satisfied, the weight would be .7. When the second rule was satisfied, the result would be .7 + (1 − .7) * .9 = .97. This scoring function, therefore, gives a bonus for accumulating evidence. This example assumes the weights are all positive. If there is both positive and negative evidence, the scores for positive and negative evidence for a hypothesis are computed separately, and subsequently added.

In summary, scoring functions are far from optimal, but in many cases can be successfully applied. Several systems, such as EMYCIN or PROSPECTOR, have used scoring functions with good results in application systems. Other systems, such as OPS, do not use scoring functions. One of the major goals of scoring functions is to resolve conflicting evidence, and to gauge the cumulative effect of individual pieces of evidence. There are also major dangers in using these functions including inaccuracies in the propagated weights and difficulties in explaining how a propagated weight is derived and explaining how to modify the knowledge base to correct a propagated weight. Since our experience indicates that the latter is of greatest concern, we have opted for the simple maximum absolute value scoring function. This scoring scheme has its defects, such as limitations in absorbing conflicting evidence, but it is easy to intuitively understand how the scoring was performed for a particular case. For many applications, it is advantageous to rely as little as possible on scoring functions.

It should be noted that not all production rule schemes which use confidence measures explicitly state the confidence range for the hypotheses appearing in the left hand side of a (HH) rule. Most production rule systems only state that the hypothesis has been confirmed, which implies that a default threshold on the confidence weight for the hypothesis has been surpassed. For example:

$$F(FCWS,T) \text{ \& } H(FLOOD,.2:1) \rightarrow H(WAIT,.9)$$

becomes

$$F(FCWS,T) \text{ \& } H(FLOOD,T) \rightarrow H(WAIT,.9)$$

While this scheme may simplify the specification of the rules, the scoring becomes more complicated, introducing another potential source of difficulty in weight propagation. As stated above, this type of scheme typically requires some minimum threshold on the confidence to be exceeded (such as .2), before the hypothesis *FLOOD* would be satisfied for this rule. If more than one hypothesis appears in the left side of rules of this type, the minimum weight of these hypotheses is used. This is a heuristic analogous to stating that the actual confidence is at least as high as the weakest link in a chain. The confidence measure appearing in the right hand part of the rule is multiplied by the derived lefthand-side weight, before the the general type of scoring function is applied. In our example, if the current weight of *FLOOD* is .8, the weight for this rule would be .8 * .9 = .72. The CF value of .72 would then be used in a MYCIN-like scoring function.

We favor the explicit expression of a confidence range for hypotheses appearing in the lefthand side of production rules. This greatly simplifies the scoring function requirements. More importantly, this should result in much more predictable and stable performance characteristics for an expert model. As an additional benefit, many of the approaches we have described here should be much easier to implement because recursion and backtracking are generally unnecessary.

If all findings were reported simultaneously and our interest was strictly in classification, we could use a very simple control strategy. First, after all findings have been given to us, the system determines if any other findings are directly implied by the FF rules. Then the simple FH rules are invoked and the ordered HH rules are processed. Because the rules are ordered, they are processed in one pass. However, we may want to build a system for which all findings are not received at once and appropriate questions must be asked. We will need a strategy for asking questions.

4.4.2 Asking Questions

As the designer of a model, one would like to somehow give the system a minimum amount of information and hope that a general strategy could be programmed so that the computer would always choose the *best next* question to ask of the user. Some might like to paint the most optimistic picture, with the computer playing the role of Sherlock Holmes, brilliantly gathering evidence, and, while everyone else is still in the dark, stopping and announcing a decision. In the real world, the expert rarely operates in this manner. In some situations a questioning strategy is not important. All information may be received at the same time, as happened in our interpretation of laboratory instrument data which is described in Chapter 5. Or else there may be a well-defined sequence or order in which the expert gathers information. For example, in medicine, experience and a tradition of systematic procedures has resulted in the patient history almost always preceding the physical examination and laboratory testing in non-emergency cases. One cannot give a single optimal strategy for asking questions. Rather, the quality of questioning will very much depend on whether the questions have been clearly organized ahead of time. If all questions are organized to produce yes/no findings, with no further structure, the result will be that for many applications no questioning strategy will work well. The system will be forced to ask too many questions, since it will have no indication of relations among them, such as checklists that allow a naturally related group of questions to be answered at the same time. One of the keys to having a good questioning strategy is to give the questions as much structure as possible. The questions should be grouped together according to common topics. Maximum use can be made of questionnaires which in their local context will force any strategy to ask the questions within the questionnaire in a pre-specified order. Very simple rules, such as the FF rules, can force local branching by topic within the questionnaire. This should work well for many applications, particularly those that are specialized, covering relatively circumscribed subject matter. It is not uncommon to design an expert model that consists of a single questionnaire.

Before we consider a strategy for asking questions, we should examine how a new finding is interpreted. With the representation described in previous sections, we have a simple scheme for classification when all findings are received together. We simply process the rules in the following sequence: FF rules, FH rules, HH rules. Because the HH rules are ordered, they can be processed in a single pass.

With an interactive expert system, findings are received sequentially. The system will ask questions, incorporate the results of questions into the working memory, and then hopefully ask intelligent questions based on the total results received. Assuming we receive a new finding, how can we add this information to the previous information and then continue with the next question? The simplest approach is to reprocess all findings again, as if they had just been received together. In the EXPERT representation, this is quite efficient because the rules are ordered. If we use question structures such as questionnaires, we know that except for the FF rules (the simple branching rules), no further processing of rules is necessary until the questionnaire is completed. The FH rules do not propagate among themselves so that only the rules which include the new findings need be processed. The HH rules, however, are reprocessed each time a new batch of findings is received.

As a model grows very large, even with a rich structure of question orderings a dynamic question selection strategy may become more important. Question selection is highly related to the control procedures for classification. There are two principal general control strategies which have been used both in classification and question selection: goal-directed and data-driven control. Usually goal-directed control is used in conjunction with backward chaining while data-driven questioning is used with forward chaining. With goal-directed control we focus our attention on a set of hypotheses and ask questions related to that set of hypotheses. We may continue to ask questions until no questions remain for the goal hypotheses. Imagine for example that we set our goal as determining whether there is a battery problem. We may therefore assume there is a battery problem and examine the rules concluding a battery problem and ask the questions which would satisfy these rules.

Alternatively, we could start with no assumptions and simply see where our evidence leads us. Thus when we receive a finding we would expand our goals to those which are potentially supported by the evidence. While goal-directed questioning and control can work quite well in highly circumscribed applications, there is a tendency to consider and chain too many hypotheses when the knowledge base is large.

For the representation we have described, the situation is somewhat different than either of these pure strategies. If we view backward and forward chaining as a form of dynamic ordering of the rules for the appropriate context, then the control strategy for the EXPERT representation need not be concerned about this step since the rules are ordered. The control strategy will be basically data-driven, but we can afford to consider all goals because only one pass is required.

Let us assume that we are searching for a single good question to ask. The presentation of the question to the user may include other questions because of structures such as the questionnaire. However, for purposes of the control strategy let's assume we are looking for a single finding. Further, our control strategy will examine *every* production rule in the system that has not been logically ruled out. Because rules are often composed of findings which are absolutely true or false, many rules may not be satisfiable for the given case. For example, if SCRNK is false then the following rule cannot be satisfied.

$$F(SCRNK,T) \ \& \ F(DIM,T) \rightarrow H(BATD,.7)$$

A finding which is unknown and has not been asked is a candidate for questioning. If SCRNK had in fact been true, but DIM had not been asked, then DIM is a candidate for questioning. As each potentially satisfiable rule which contains unknown findings is examined, the control system tries to determine whether the finding might prove to be *better* than its current *best* question. Many factors may be considered in determining why a question is better than another. Expert systems have been less successful with question strategies than they have been for classification. The goal of maintaining a coherent and sensible dialog has not always been achieved. Here are some of the intuitive considerations that are used by EXPERT and other systems to determine during a session whether one question is a better candidate than another.

1. Ask the least costly question. The questions can be assigned cost or risk measures and the least costly questions ought to be asked before the more expensive questions. Since precise cost or risk in the literal sense is very difficult to assign, cost is usually taken to be an approximate ordering heuristic.
2. Give preference to those questions which affect the currently highest-weighted hypotheses. Select questions which appear in production rules which conclude those hypotheses with the highest current confidence.
3. Consider only those hypotheses which are related to a currently reported finding.

4. Consider only those findings which can potentially increase or decrease the current ranking of a hypothesis by some specified threshold.

5. Terminate questioning if the confidence in any hypothesis exceeds a predetermined threshold. This type of termination strategy is not used often, because one would usually want to ask more questions rather than stop early with a possibly wrong answer. This is particularly true for a system which has sequential reasoning capabilities that immediately allow the user to see the current ranking of conclusions.

The relative importance of each of these factors is a design decision. In our experience, we have infrequently used factors 1 and 5. Factor 2 allows the system to focus its questioning on the most likely possibilities. This is particularly important for a system that has great breadth, i.e. covers many hypotheses. It is also useful in the early stages of questioning, when the system is looking for direction. One might be concerned by an overemphasis on the highest weighted hypotheses to the detriment of other possibilities. However, factor 3 will allow questioning to continue for other possibilities. Depending on the requirements established for marking hypotheses related to the evidence, a procedure for implementing factor 3 can vary from simple to complex. For example, consider the following rule:

$$F(X,T) \ \& \ H(A,.1:1) \ \rightarrow \ H(B,.9)$$

If X has been reported false, this rule would not interest us in A or B. If finding X has been reported to be true, we would probably like to consider (mark) hypotheses A and B if they are unknown. If X is unknown, our interest in X would depend on the status of A or B. In general, if a rule has not been proven false, and any single element has been satisfied or has been marked of interest by another rule, we may want to mark all elements of that rule to be of interest. The tracing of these relationships can get complex if we maintain rigorous logical standards for marking hypotheses to be of interest. We can simplify greatly the marking procedure by relaxing our standards. For example, instead of tracing the relationships between all reported findings and hypotheses, we might use "context" information to simply mark all hypotheses and findings appearing within the satisfied context. For example, using the EXPERT syntax, when the "*IF" rules are satisfied, mark every element which appears in a rule within the "*THEN" section (see section 4.3.3.3).

For factor 4, if the maximum absolute value scoring function is used, a simple test can be carried out to see if the confidence of the rule in question is greater than the current confidence in the hypothesis. A rule can be ignored if it's confidence measure is less than the current value for the hypothesis. If other scoring functions are used, a limit could be specified on how close the confidence implied by the rule must be to the current confidence. For example, if the confidence must be within .2, a rule which yields a confidence in hypothesis A of .1, when the current confidence of A is .8 will not be considered.

4.5   Explanations of Decisions

There are two very different types of users who might be interested in an explanation of a decision. These are the model designer and the end user of the system. Researchers in the field of expert systems have disagreed on whether an explanation suitable for the model designer is adequate for the end user of the system. For the model designer, most people generally agree that for a production rule system, displaying the set of rules which were satisfied for a particular hypothesis is the most direct form of explanation. When confidence measures are used, a complex scoring function can make it difficult to provide an easily understood explanation of how the final numerical ranking of a hypothesis was achieved. When confidence measures are not used, or when a simple scoring function such as the maximum (absolute) value is applied, a form of explanation is provided by citing the single rule which was used to conclude the hypothesis in question. If this rule also refers to other hypotheses, then the related hypotheses can be traced and the rules for these hypotheses can be cited. The following is an example of this type of explanation given for the sample session of the auto repair model. In this case the user requests information about the hypothesis BATD, *Battery Discharged*.

:hypo(BATD)

Battery Discharged was set by rule 7 in this manner:

   If: Starter Data: No Cranking
            Simple Checks: Headlights Are Dim
   Then: *Battery Discharged* (0.9).

This type of information is particularly valuable to the model designer in evaluating the current logic of a model and making modifications which may lead to improved performance.

Some designers of expert systems believe that this type of explanation is inadequate for the end user and that such forms of explanation are likely to be too stilted and not logically satisfying to the users of the system. One approach to explanation that covers some of these objections is to augment the conclusions with statements that are more informative than just stating a conclusion. Hypotheses may be any form of statement that is appropriate including interpretations, advice, and explanations. The model designer may be able to anticipate those explanations which are appropriate for a given hypothesis. For example, in the auto repair model, instead of separating the conclusions into two categories (diagnoses and treatments), a general interpretation may be given which is somewhat explanatory. Such a statement might be of the form:

Wait 10 minutes to depress the accelerator to the floor because the car is flooded.

Various prototypical models will be discussed in Chapter 5. An important point that we wish to emphasize is that careful phrasing of the conclusions will often provide an implicit form of explanation for prototypical models.

4.6   Useful System Design Attributes

We now have the basic building blocks of an expert system designed around the classification model. Many problems will fit into this framework. Even if a problem does not completely fit the classification model, it is likely that a general production rule scheme will be fundamental to the design of the model. Although one can build a model specific to an application using the concepts which have been described, it would be advantageous to have a general facility for expressing these types of models. The model design process will probably be a dynamic process with many changes being constantly suggested by the expert. One would like to have a degree of stability in the model design process, with the model designer knowing that many changes can be made to the model without a major reprogramming effort. A general system which is designed for the classification model can then be tailored to the specific application. At some point it may be necessary to write code specific to the problem. However, one may still take advantage of the basic classification model for the initial design and prototype development phase.

An expert system based on the classification model can be programmed in most higher level languages. The EXPERT system, which is an implementation of these ideas, was coded in FORTRAN. For schemes which require extensive list processing, backtracking, or recursion, a language such as LISP is preferable. It is quite possible, however, to think of alternative representations such as those described here, which can be implemented in languages which have fewer capabilities than LISP. The implementation will require that the model be translated into some intermediate representation which then can be handled more efficiently and easily by the actual running program. This intermediate representation will organize the model in an efficient way, setting up the necessary lists, arrays and pointers. For example, the process of creating and running an EXPERT model is similar to writing and running a conventional computer program. A standard text editor is used to create a file which will contain statements in a special purpose programming language used to describe a model. This is the same language used to illustrate the auto repair model. A model is checked for syntactic errors and translated into an efficient internal representation by the compiler program. The model may then be executed, and cases may be entered for consultation.

Processing the model as a programming language somewhat simplifies the implementation of the system. One can easily change the model by editing the file, and one has the opportunity to check for errors and to translate the model into an efficient representation. This is the approach we prefer. There are some disadvantages. Permanent changes to the model will require that the complete model be compiled again. If the model designer expresses the model in a language such a LISP which is usually interpreted, the compilation process may be bypassed. An alternative to a special purpose language for expressing a model is to design a interactive knowledge acquisition system. This system would prompt the model designer to supply the necessary information to define or modify a hypothesis, finding, decision rule, or any other construct necessary to describe the model. If the interface program between the model designer and the model description is well done, the task of designing a model may be simplified, particularly for a relatively complex language. It is no easy task, however, to design such a system. It is not unheard of to see a knowledge acquisition interface which is difficult and cumbersome to use, defeating the whole purpose of such a system. We prefer the specialized language approach which minimizes the need for a knowledge acquisition interface. Of course, a poorly designed language will also slow down the knowledge acquisition process.

There are several additional capabilities of a general system which may prove useful in designing models. These include:

- Accepting volunteered information from the user, including both corrections of previous responses and new information.
- Sequential processing of all information. For example, one may ask for the system's conclusions at any point in a consultation session.
- Facilities for storing and retrieving cases.
- Tracing facilities for a case. An automatic trace for a case would monitor a change in status of the working memory, such as a change in a finding or hypothesis value or an indication that a rule was invoked and was (or was not) satisfied. An alternative to automatic tracing is a facility for inquiring as to the the status of any hypothesis, finding or rule in the system.
- Identification of missing findings that would strengthen conclusions. For example, some questions may have been asked and the user did not have the information. The rules which have not been satisfied are examined to see if they could change the current conclusions.

- Determination of inconsistencies. Some inconsistencies may be detected automatically, such as rules satisfied with both positive and negative confidence for the same hypothesis. While this is not unusual for real problems, the user should be informed of these inconsistencies. Other inconsistencies may be anticipated by the application system designer. He may write rules to point out inconsistencies in the reporting of findings. For example, in a car repair model, some findings may be inconsistent with carbureted versus fuel injected engines.

We have now seen how one may go about representing and running a model in an expert system. In Chapter 5, we'll examine some thoughts on making good choices in describing a specific model application.

5

How to Build a Practical System

5.1 The Art of Designing an Expert Model

Acquiring knowledge from an expert is a gradual process that can stretch over weeks, months, or even years. Prototype models of increasing complexity are built and refined until the system performs at an acceptable level. The design of an expert system is an art and the model designer must be be willing to constantly adapt to a dynamically changing environment. Of fundamental importance is the selection of a realistic application. One's chances of success are greatly enhanced if the problem fits some class of problems for which solutions have been found, such as classification problems. One must also consider very carefully whether the proposed application system would be of significant value if implemented. While this may seem obvious, many systems have progressed to relatively advanced stages, before it is recognized that there is no market for the application system.

The most difficult part of developing a system is the design effort in its early stages. One of the distinguishing characteristics of the design of an expert system is that one is constantly aiming at a moving target. While many software systems can be designed with firm specifications, expert systems usually cannot be readily specified before the system is actually built. A mistake of many novice expert system designers is to dwell on the preliminary knowledge-acquisition phase of system development. In some projects long periods of time have been spent on expert interviews and literature searches without any clear direction on how to proceed. Somehow the designers feel that by continually gathering information, one will achieve insight into the eventual solution of the problem. Usually, just the opposite will occur. One will be overwhelmed with the mass of information, without getting closer to the development of the expert system. The human expert may not always be very helpful in giving direction to the project. He may be an expert in a specialized field of knowledge, but is usually not an expert in designing a system. Worse yet, while the expert is extremely knowledgeable in his field, he may not be able to formalize his reasoning. The expert may to some degree be able to explain his reasoning for a particular case, but the generalization of the reasoning to cover many cases will prove far more difficult.

The single most important piece of advice that one can give a model designer is to *build a prototype model as soon as possible*. Because expert reasoning problems are frequently poorly specified, one needs to have something concrete to view and "lay hands on." It is particularly important for the expert to see something running early. A running program is worth thousands of words from an unformalized interview with the expert. The initial prototype may be crude, will certainly be incomplete, and may contain inaccuracies, but at least it provides a focused point of departure from which the expert can make his suggestions. One of the sacrifices one has to make in building expert systems, is to be constantly told about the weaknesses and flaws in the system that is being designed. One has to expect that major revisions will have to be made in the system, particularly in the early stages. It is sometimes amazing to watch the pace of useful knowledge acquisition accelerate once a prototype model has been built. Instead of abstract suggestions from the expert and the model designers, one has a much more limited basis for converging to a practical system. And to those who point out that such constraints will fatally bias or limit the ultimate system design, the answer is that one must always keep an open mind to the need for making drastic changes of representation, and must be willing to put in the effort and resources to carry out these changes if they prove necessary.

Much of the burden of the model design will fall on the knowledge engineer who extracts this knowledge in a form suitable for computer representation. While this task will always be an art, there are some rough guidelines that we have found valuable in pointing the model designer and the expert in the right direction for building the initial prototype, particularly if the model can be posed as a classification model.

1. Design a model by focusing first on a small set of hypotheses and include in a first prototype only those findings that are most predictive of these hypotheses.
2. Identify clusters of findings that are most discriminating.
3. In the decision rules, combine the smallest number of findings necessary to confirm or discriminate among hypotheses. Increase the number of conjunctive rules when the resulting rule will significantly increase the power of the system to confirm or deny the conclusion. There is potentially a large number of combinations of findings. A model should contain the smallest number of these that are sufficiently specific to confirm, deny, and discriminate.
4. Include findings that may not be strongly predictive or discriminatory on their own, but which can significantly improve the quality of decisions by setting a context or focus of attention for the decision-making process. The prototype system at this stage becomes more realistic.

5. Determine whether abstractions can be made, for example, some production rules can be stated as being satisfied if *n or more* out of a list of findings are satisfied, e.g. 2 out of 10. Such rules may prove useful in expressing combined effects for relatively independent sets of findings that, on their own, contribute marginally to an inference.

6. See whether additional intermediate hypotheses can be introduced to simplify reasoning. There may be many different combinations of findings that enter into rules for each of a set of hypotheses, such as H_1, H_2, and H_3. Using conjunctions of intermediate hypotheses (H_1 & H_2 & H_3) will reduce the proliferation of combinations of production rules. It is also often practical to determine the hypothesis H_1 independently and use the confirmation or denial of the H_1 for further reasoning, instead of always reasoning with simple findings (the FH rules).

7. Test the model on a data base of cases. Consider those cases that yield inaccurate conclusions, and determine how the model may be changed to correct the errors. Revise the model and examine the conclusions for these cases and the effect of these revisions on the other saved cases.

5.2 A Microprocessor-based System for Laboratory Instrumentation

In this section we will look at an expert system that has been fully developed and implemented as a commercial application: the Serum Protein Diagnostic Program. While many problems may prove more complex than the interpretations necessary for this application, the problem presented here does illustrate all the steps necessary to take an expert system from beginning to end. The problem we will examine, that of the interpretation of serum protein electrophoresis, is a pure classification problem which is easily posed in the format we have previously described. In order to achieve commercial applicability, the model was later translated, by an automatic process, into microprocessor assembly language. However, the greater part of the development of the model proceeded as for any other model. Only after we were satisfied with the model's performance did the translation take place.

As we have observed in Chapter 3, most knowledge-based medical consultation systems developed during the 1970s were relatively large-scale experimental prototypes. Their advice on diagnostic and treatment problems typically involved approximate reasoning over a space of many interrelated hypotheses, supported by hundreds of observations linked to them by various types of reasoning rules. By adopting symbolic reasoning methods with more powerful representations than the traditional mathematical decision-making schemes, these knowledge-based systems produced results that were generally easier to analyze, explain and update than those from the more statistical systems. Human engineering features were often stressed as an important means of enhancing the interaction with the expert systems. Successful clinical experience with many of these systems has been reported in pilot demonstration projects, yet none is in routine clinical use at present. Both technical and social factors contribute to the difficulties of introducing expert systems into the everyday practice of medicine. One oft-cited technical factor is the slow rate of manual data entry required by most of the larger systems. This problem is minimized for applications where most of the data can be read directly off a clinical instrument, and only few items need be entered manually. The commercial availability and use of automated electrocardiogram- interpretation programs (using traditional algorithmic techniques) supports this point. Regardless of the methods used in constructing a knowledge base or its complexity, instrument-derived interpretations are more likely to be accepted because they can be seen as extensions of the instrument. And since many advanced medical instruments are already microprocessor controlled, it may be possible to add an interpretive module that enhances the performance of such a device at relatively little extra cost.

Using the classification model approach to designing an expert system, we were able to accelerate the development of interpretive software for a widely used laboratory instrument, the scanning densitometer. This was accomplished by automatically producing a computer translation of an expert model for serum protein electrophoresis interpretation, developed on a mainframe computer, into a microprocessor assembly-language version. Such translation methods have been generalized so that this process can be repeated for EXPERT models in other domains, with a few restrictions on the types of knowledge structures used.

The process of model design and transfer that we used in developing the microprocessor-based expert model for serum protein electrophoresis interpretation involved the following tasks:

- specification of the knowledge base using EXPERT
- empirical testing with several hundred cases
- refinement of the knowledge base by the expert
- further cycle of testing with additional cases, and review by independent experts
- test of the final model on the large machine
- automatic translation of the EXPERT model to a specialized program and a microprocessor assembly language program
- interfacing of the assembly language model with the instrument.

This last step requires detailed knowledge of the instrument. In this application the manufacturer, Helena Laboratories, interfaced the interpretive program to the existing program for printing instrument readings.

Figure 5.1 is an example of the interpretive analysis of the instrument. The output of the instrument is displayed as a waveform with five peaks. The

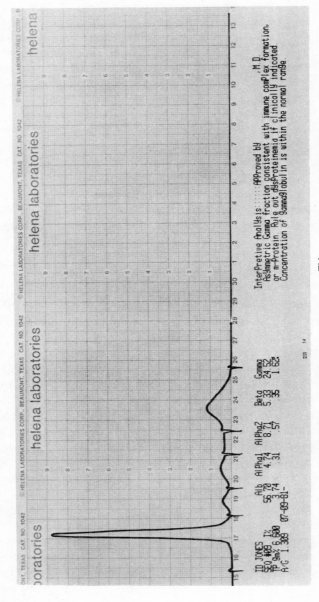

Figure 5.1: Output of the CliniScan ™ with interpretive analysis

instrument lists two series of numbers for each of the five "fractions." The numbers used in the interpretation are the second series of numbers which represent the measured areas under each peak. These numbers were available directly from the instrument, and the task was to interpret the meaning of these numbers when correlated with several other items. An example of the production rules used to form this interpretation is given in Figure 5.2. This example illustrates two production rules. The first rule is an interpretation of a specific abnormal pattern. The second rule points out an interesting form of reasoning on the part of our expert. In general, the expert does not indicate normal values in his interpretation. He tries to determine whether all findings are normal, or he identifies the abnormalities. However, in the presence of an abnormality regarding the "gamma" tracing, the expert wants the program to indicate that the level of concentration is normal.

If: The the tracing pattern is "asymmetric gamma" and
        the patient's age is greater than 25 years

Then: *Asymmetric Gamma fraction consistent with immune complex formation, or m-protein. Rule out dysproteinemia if clinically indicated.*

If: The the tracing pattern is "asymmetric gamma" and
        the gamma quantity is normal (correlated with age)

Then: *Concentration of gammaglobulin is within the normal range.*

In section 5.2.1 we will discuss how these interpretive statements can be composed so that they are presented in an organized manner. With slight modifications to the simple production-rule representation, we can place an ordering on the statements so that some statements are displayed before others, and redundancy is eliminated by introducing rules which manage the display of the interpretive statements.

An abstracted version of the EXPERT model for the serum protein interpretation is given at the end of this section. The form of this model is quite similar to that of the auto repair model in Chapter 4. The most significant restriction on the type of production rules used in the model was to limit the use of confidence measures to three values indicating confirmation, denial, or unknown status. In this application, these three values are sufficient to capture the expert knowledge and yet give the model designer enough freedom to confirm and deny hypotheses at the proper point in the analysis. Because many of the conclusions are not mutually exclusive, a decision tree is inappropriate to this problem. One needs a mechanism to experiment with various rules, modifying and adding rules as necessary, and then reaching the steady state of a satisfactory model. By clever ordering of the rules, the model is simplified. Note that at the very end of the model a whole series of hypotheses are denied. These hypotheses are important in the intermediate

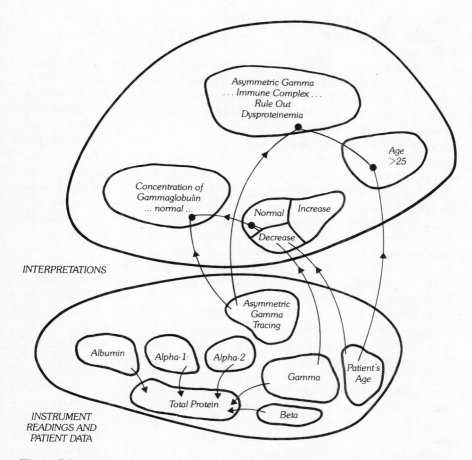

Figure 5.2:   Excerpt of the expert logic from the interpretive model in the CliniScan™

stages of reasoning, but are to a large extent redundant in the interpretive output, once more specific conclusions have been established. After the developmental process is completed and a satisfactory model has been designed, it is relatively easy to translate the required reasoning rules into direct code.

In applications of this type, it should be noted that the strategy of questioning is not a significant task because most of the information will be obtained directly from the instrument. In building the EXPERT model we simulated this situation by entering the values of certain key features of the instrument signal (Figure 5.1) that are currently given as a digital output by the instrument. These features include peaks of the waveform and areas under certain segments of the waveform. A few items (patient identification, age, and some waveform features that are more easily scanned by the technician) are entered manually.

The serum protein electrophoresis model required several stages of refinement over a period of six months, with the aid of one principal expert and comments and suggestions from the independent experts. We began with a relatively small and simple model, having 10 conclusions and a production rule for each. After the first cycle of revision we had about 25 conclusions and 50 rules which included many rules for handling interactions among the hypotheses. The current model has 38 conclusions and 82 production rules. Its performance on the initial 256 test cases covering the full spectrum of conclusions is 100% acceptable to our experts. They expect differences of opinion on the amount of detail included in the present set of conclusions, but feel that covering infrequently found problems in a more detailed manner would detract from a model that is to be disseminated widely. An option for allowing users to add a written record of their own opinions on such unusual cases has been provided in the final microprocessor implementation.

The completed microprocessor (in this case a Motorola 6809) version of the interpretative serum protein electrophoresis model may not look much different than if it had been hand coded directly in the assembly language of the microprocessor, or translated from an algorithmic language. There is, nevertheless, a fundamental difference. With this approach, new versions of the microprocessor program can be rapidly produced from an EXPERT model in response to changes suggested by the experts or to changes resulting from future empirical analysis and clinical tests in the field. In contrast, considerable effort would usually be required to recode directly on a microprocessor. Besides, the original expert-derived model is also very different from one produced by more traditional methods. The conclusions and intermediate hypotheses were developed to model the expert. The large amount of experimentation that took place with the model as it went through its cycles of testing and modification could only be carried out on a larger system, with adequate facilities for analyzing many cases and knowledge-engineering tools for changing the model. A common mistake of many standard computer-based techniques in interpreting the results of laboratory tests is that they reach conclusions that are overly specific given the nature of the supporting data.

This microprocessor-based project is an illustration of the requirements needed in taking an expert system from an early developmental phase to actual implementation and use in the real world. Such applications can lead to the increasing acceptance of expert systems in medicine and other domains where similar problems can be found. A particularly powerful and promising approach to implementing software for wide dissemination is to develop the expert system on a large machine having all the necessary tools for system development. The completed knowledge base and a simplified control strategy may then be translated into a lower-level form which may be run on a microprocessor. The model for serum protein interpretation is similar to other classification models, showing very little difference from the auto repair prototype described in Chapter 4.

**hypotheses
*taxonomy

Here is the list of conclusions the instrument can reach. A "+" indicates a continuation.

UNUS	Unusual pattern. See accompanying description.
MYEL	M-spike present. Rule out dysproteinemia; Suggest+ serum and urine immunoelectrophoresis.
DG	Decreased gamma globulin.
IG1	Increased gamma globulin.
CIR	Electrophoretic pattern suggests cirrhosis of the liver.
SINFL	Electrophoretic pattern suggests subacute inflammation.
IA1D	Increased alpha 1 globulin and Hypoalbuminemia+ consistent with acute or subacute inflammation.
IA1	Increased alpha 1 globulin—consistent with acute+ or subacute inflammation.
IA2D	Increased alpha 2 globulin and Hypoalbuminemia—+ consistent with acute or subacute inflammation.
IA2	Increased alpha 2 globulin—consistent with acute or+ subacute inflammation.
IGD	Polyclonal increase in gamma globulin and+ Hypoalbuminemia—consistent with chronic inflammation+ or infection.
IG	Polyclonal increase in gamma globulin—consistent+ with chronic inflammation or infection.
IALB	Increased albumin—consistent with dehydration.
DALB	Hypoalbuminemia.
NORM	Normal electrophoretic pattern.

•
•
•

2 useful intermediate conclusions

*Intermediate hypotheses
GPAT   Gamma pattern (spike or asymmetric)
DPAT   Disease Pattern

A simple list indicating the order in which conclusions will be printed if they are appropriate; see section 5.2.1

*print control
UNUS,MYEL,DG,IG1,CIR,SINFL . . .
IALD,IG1,IA2D,IA2,IGD,IG,IALB,DALB,NORM

Here are the findings; all except age and some tracing characteristics will come directly from the instrument.

**findings

This is the order in which the questions will be asked.

*begin questionnaire
*numerical
AGE   Age:
*numerical
AGEM   Age (months):
*numerical
ALB   Albumin gm/dl
*numerical
A1   Alpha-1 gm/dl
*numerical
A2   Alpha-2 gm/dl
*numerical
BETA   Beta gm/dl
*numerical
GAMMA Gamma gm/dl
*multiple choice

Tracing characteristics:
BRIDG   Beta-gamma bridge
SPIKE   M-spike
AGAMF Asymmetrical gamma
PSPK   Possible m-spike
UNUF   Unusual pattern
*checklist
Other tracing characteristics:
FIBET   Fibrinogen beta spike
*end questionnaire

One simple computation is needed; this too is already available from the microprocessor.

*functional findings
TP = ALB + A1 + A2 + BETA + GAMMA
**RULES

Just a simple rule not to ask about age in months if the patient is older than 1 year.

*FF
F(AGE,1:999) → F(AGEM,F)

Here the important production rules begin.

*FH

F(UNUF,T) → H(UNUS,.9)

.
.
.

*HH

If there is no indication of an unusual pattern, evaluate the following rules.

*IF
H(UNUS,-1:0)
*THEN
Tracing patterns
F(SPIKE,T) → H(MYEL,.9)
F(BRIDG,T) → H(CIR,.9)

.
.
.

Rules for intermediate hypotheses

[1:H(MYEL,.9:1),F(PSPK,T),F(AGAMF,T)] → H(GPAT,.9)

.
.
.

Independent interpretation of each instrument reading

F(ALB,0:2.89) → H(DALB,.9)
F(ALB,5.31:999) → H(IALB,.9)

•
•
•

Gamma values are correlated with age

F(AGEM,0:4.99)&F(GAMMA,0:.19) → H(DG,.9)

•
•
•

Patterns for interpretation

F(ALB,3.2:5.3)&F(A1,.1:.4)&F(A2,.4:1)&F(BETA,.5:1.1) +
&H(DG,-1:0)&H(IG,-1:0) → H(NORM,.9)

F(ALB,0:2.49)&F(A1,0:.14)&F(A2,0:.50)&F(BETA,1.11:999) +
&H(IG,.9:1) → H(CIR,.9)

•
•
•

Format rules to eliminate redundancy in the final output

[1:H(DPAT,.9:1),H(IA1D,.9:1)] → H(IA1,-1)
[1:H(DPAT,.9:1),H(IA2D,.9:1)] → H(IA2,-1)
[1:H(CIR,.9:1),H(SINFL,.9:1),H(IGD,.9:1),H(IG1,.9:1)]→H(IG,-1)
[1:H(DPAT,.9:1),H(SINFL,.9:1)]→ H(DALB,-1)

•
•
•

*END

5.2.1 COMPOSING INTERPRETATIONS OF RESULTS

In the auto repair prototype model, we saw that the conclusions for a particular case were listed in order of decreasing levels of confidence and with the
confidence measure assigned by the program. There are several alternative
schemes for presenting the conclusions of the program to the user. While it
may seem that the method of presentation of conclusions is merely a matter of
aesthetics, the display of the conclusions is at the heart of the model design

process. Since the goal of the model design is to reach the correct conclusions, the display and formatting of these conclusions will affect most phases of model design. For the auto repair model, we used the simplest scheme. Conclusions which were confirmed were listed in weighted order. This required the least effort on the part of the model designer.

In the Serum Protein Diagnostic Program an alternative but related scheme was used. In this case the model designer did not think of the conclusions as absolute endpoints that should be ranked in order of highest to lowest confidence. Instead of conclusions separated into categories such as diagnoses or treatments, there is only one category: a class of interpretations. The task of the system is to produce a set of sentences that describe the interpretation of a case. The interpretations are not ranked; they are either appropriate or inappropriate to the case. While internally one may use confidence measures, it is understood by the model designer that when it is time to display the conclusions the confidence measure will not be displayed. The differences among positive confidence measures for interpretive displays are not considered significant. This is not just a question of the display of output. Rather, our understanding of the conclusions is different. We do not expect interpretations to be competing hypotheses. The interpretation is a description of the appropriate statements for a specific case.

In order to make sensible statements, we must have some means of ordering the statements. For interpretations, this order must be specified by the model designer. In the Serum Protein Diagnostic Program, a very simple device was used which has proven effective for many different applications. This scheme requires that the model designer specify a single ordered list of hypotheses which indicates the order of the display of conclusions. The most important hypotheses may be printed first and others may not be printed at all.

For example, suppose we specify ordering of some of the auto repair model hypotheses in terms of the following list of mnemonics: *FLOOD, CHOKE, EMPTY, CAB, BATD, STRTR.* For any given case, only those hypotheses which are confirmed and are found on this list will be printed. They will be printed in this order: *FLOOD before CHOKE, CHOKE before EMPTY,* etc.

While this scheme may appear to be very simple, it has proven its value many times. We used this approach in the Serum Protein Diagnostic Program with excellent results. The model designer must, however, think about the model with the potential interpretations in mind. The conclusions are the interpretive statements which can be made by the program. The rules must be written to capture this type of analysis, to confirm hypotheses which are appropriate interpretations. An effective technique for controlling the final printout is to include rules in the model with the sole purpose of suppressing the display of conclusions which are useful in the reasoning but are covered by other statements. This will help in simplifying the rule set. These display techniques for conclusions were used in the Serum Protein Diagnostic Program, as shown in the model description. The confidence measures are used only to confirm and deny the hypotheses. The statements labeled *print control* are used to specify the order of printing the interpretive statements. The final set of rules suppresses the printout of conclusions that are covered by other more significant conclusions.

5.3   Decision-Table Models in Expert Systems

Production rules are relatively simple structures that can be used to describe an expert's reasoning rules. Most such rules are expressed as a series of conjunctions: A & B & C imply D. A relatively simple extension to basic production rules would allow a choice among several items which may imply a conclusion. For example, any two or more of the set (A, B, C) imply D. Rules of this form are the basis for a type of decision table which we, together with our collaborators, have used successfully in building an expert system for diagnosing rheumatic diseases. The model for each disease can be described in terms of a table of diagnostic criteria. Instead of using numerical confidence measures in the rules, the model designer expresses a rule by indicating that a diagnosis is believed at either possible, probable, or definite levels. This type of categorization is a somewhat restricted but highly understandable means of expressing expert knowledge and managing a knowledge base. The scheme which will be presented in this section is but one of many which could be described to express expert knowledge in a uniform manner. The common basis for understanding how to use this knowledge in an expert system is that these models map into forms of production rules and classification models.

The application is for an expert system under development to represent expert medical knowledge for a set of rheumatological diseases. The initial direction of this work was in providing assistance in the differential diagnosis of seven related diseases which constitute a particularly complex and confusing problem area. The differential diagnosis in this area is quite difficult; even the experts may disagree about some of the diagnoses, and clear criteria to objectively confirm the diagnoses have been lacking. The model has since been greatly enhanced to describe many additional diseases beyond the original set of diseases.

Throughout the development of this expert model, performance has been evaluated by testing against a data base of clinical cases which includes the correct diagnosis for each case. In this application, a correct diagnosis was decided by an agreement of at least two out of three experts. Since the initial prototype system, consisting of 18 observations and 35 rules, was produced, the model has undergone many cycles of testing and revision. This incremental process resulted in the expansion of the model to include 150 observations, of which several observations were combined by rules to reach intermediate conclusions, and a total of 147 rules. The model has been critiqued by an external panel of experts, and a review of performance has shown the model to achieve diagnostic accuracy in 94% of 145 clinical cases. The model has since been expanded to 26 diseases, with over 1000 production rules and 800 observations, and runs on several machines including a relatively inexpensive Motorola 68000–based microcomputer.

As is fairly typical for medical models, the questions are organized by related topics. The user responds by typing numbers to indicate the appropriate findings. The process of describing a case is akin to filling out a questionnaire with specific observations. Questioning can proceed in a thorough but reasonably efficient fashion by using the types of structures for organizing questionnaires which were described in Chapter 4. The organization of the questions is important: the design goal is to allow the user to enter information without having the system ask unnecessary questions. In this model, our initial attempt at questioning is to structure each question in a relatively rigid manner. Topics tend to proceed from the general to the specific, so that if there is no information for the general topic, the specific questions are not asked. For each case, the correct physician diagnosis is also entered. This proves valuable in evaluating the performance of a model. Editing facilities are available to review and to change the responses to questions. Cases are stored in a data base which is maintained by the system. The following example shows the typical form of data entry for a particular case. It is no different than the format for the auto repair example.

Enter Name or ID Number: **John Smith**

Enter Date of Visit: **6/22/82**

1. Extremity Findings:
   1) Arthralgia
   2) Arthritis ≤ 6 wks or non-polyarticular
   3) Chronic polyarthritis > 6 wks
   4) Erosive arthritis
   5) Deformity: subluxations or contractures
   6) Swollen hands, observed
   7) Raynaud's phenomenon
   8) Polymyalgia syndrome
   9) Synovial fluid inflammatory
   10) Subcutaneous nodules
   Checklist:
   *1,2,3,4,10

After all questions have been asked, the system provides a summary of the data for the case. Examining this summary, the expert can correct any data entry errors and, later, the case can be stored in a data base. The expert can request the system's diagnosis for a case at any time during the session. An example of the interpretative analysis output by an early version of the program is shown in Figure 5.3.* This includes the differential diagnosis, i.e. definite rheumatoid arthritis (RA) and possible systemic lupus erythematosus (SLE), followed by detailed lists of findings which provide a measure of explanation for the program's conclusions. The diagnoses are reached using a tabular scheme to represent production rules. The findings lists are obtained by matching findings from the case data to prespecified lists that are associated with each final diagnosis in the model. The lists include those findings consistent, not expected, and unknown for the diagnosis. The model for this application is a classification model using production rules. However, the model can be described in a stylized format which may simplify the knowledge acquisition process. As we shall see in Chapter 6, this type of model lends itself to performance evaluation.

INTERPRETIVE ANALYSIS

Diagnoses are considered in the categories
definite, probable, and possible.

Based on the information provided,
the differential diagnosis is
    Rheumatoid arthritis (RA)   — Definite
    Systemic lupus erythematosus (SLE) — Possible

Patient findings consistent with RA:
    Chronic polyarthritis > 6 wks
    RA factor (l.f.), titer 1: < 320
    Subcutaneous nodules
    Erosive arthritis

Patient findings not expected with RA:
    Oral/nasal mucosal ulcers

Patient findings consistent with SLE:
    Platelet count, /cu mm: $\leqslant 99999$
    Oral/nasal mucosal ulcers
    Arthritis $\leqslant 6$ wks or non-polyarticular

---

*Figure 5.3 and other examples presented from the rheumatology model are from a version of the model which has since been extensively revised and improved.

Patient findings not expected with SLE:
Erosive arthritis

Unknown findings which would support the diagnosis of SLE:
LE cells
DNA antibody (hem.)
DNA antibody (CIEP)
DNA (hem.), titer 1:
FANA
Sm antibody (imm.)

End of diagnostic consultation: 22-Jun-82.

Figure 5.3:   Sample session from rheumatology consultation program

A table of criteria, which is a specialized type of frame or prototype, is prepared for each potential diagnosis. The table consists of two parts:

- major and minor observations which are significant for reaching the diagnosis
- a set of diagnostic rules for reaching the diagnosis.

Figure 5.4 is an illustration of observations for mixed connective tissue disease.

Major Criteria	Minor Criteria
1. Swollen hands	1. Myositis, mild
2. Sclerodactyly	2. Anemia
3. Raynaud's phenomenon or esophageal hypomotility	3. Pericarditis
	4. Arthritis $\leqslant$ 6 wks
4. Myositis, severe	5. Pleuritis
5. CO diff capacity, nl: $<$ 70	6. Alopecia

Figure 5.4:   Example of major and minor observations

The second part of the table contains the diagnostic rules. In Figure 5.5, each column consists of a rule for a specific degree of certainty in the diagnosis. There are three levels of confidence: definite, probable, and possible. A diagnostic rule is a conjunction of three components taken from each row: specific numbers of majors or minor observations, requirements, and exclusions. Requirements are those combinations of observations which are necessary beyond the simple numbers of major and minor findings (although major and minor findings also may be requirements). Exclusions are those observations which rule out the diagnosis at the indicated confidence level. The three fixed confidence levels are an important attribute of the model. They substitute for complex scoring functions which can be a major difficulty in analyzing and explaining model performance. It is understood that if a definite diagnosis for a particular disease is made, even if the rules for the probable or possible diagnosis for the same disease are satisfied, the definite conclusion is appropriate. However, even if four or five rules for a diagnosis at the possible confidence level are satisfied, the diagnosis remains at the possible level of confidence. In most applications, multiple rules are described for each confidence level.

	Definite	**Probable**	**Possible**
	4 majors	2 majors	3 majors
			2 minors
Requirements	Positive RNP antibody	Positive RNP antibody	No requirement
Exclusions	Positive SM antibody	No exclusion	No exclusion

Figure 5.5:   Decision table for mixed connective tissue disease

As an example, a rule for concluding definite mixed connective tissue disease can be stated as following production rule:

If: The patient has four or more Major observations
          for mixed connective tissue disease and
          RNP antibody is positive and
          SM antibody is not positive
Then: *Conclude definite mixed connective tissue disease.*

Frame-like schemes have been used to represent medical knowledge in the PIP and CENTAUR systems, which were designed to provide diagnostic consultations in subspecialties of medicine. Unlike the tables presented here, these frames contained slots for relatively complex scoring functions that could be specialized for the evaluation of the disease frame. The tabular model is a simple type of frame representation requiring fixed types (e.g. majors, exclusions) of observations for each diagnostic conclusion that are relatively easy to understand. Also, scoring follows directly from the three confidence levels of definite, probable, or possible. In effect we have a highly modular scheme that maps directly into production rules. This tabular model can be translated directly into production rules in the usual EXPERT format.

The key idea for expressing the expert's rule in this application is to have an easily understood and uniform representation of knowledge. This is provided by the decision-table model. The strongest explanation for a model's decision from a logical point of view would seem to be the citing of decision rules which support the highest-weighted diagnosis. The factors weighing against the top choice would then be expressed by citing those rules which were satisfied for other diseases. Unfortunately, such an approach, as well as others that may involve an explanation that is satisfactory to the designer, can prove somewhat unsatisfactory to the expert user. In some cases, such detailed information may obscure rather than clarify or explain decisions in ways which are acceptable for the specialty. Often, a simpler description of supporting evidence may be more acceptable. In the rheumatology model, three lists can be created for each diagnosis, findings consistent with a conclusion, findings unexpected for a conclusion, and findings which are unknown but which would strongly support a conclusion. Once the conclusions for a particular case are determined by examining the decision rules, these lists are then examined to determine the corroborating and inconsistent evidence independently of the rules used in the reasoning.

5.4 Giving Advice for Controlling Complex Programs: An Expert System which Helps in Oil Exploration

Hart has speculated on the potential enhanced performance of expert systems having a multi-level design. He contrasts *surface-level models* with *deep models* of reasoning. Hart has described a hypothetical system for advising petroleum engineers using a multi-level approach. The surface-level model is of the production rule type, whereas the deep model is a purely mathematical description of an oil reservoir expressed as a set of partial differential equations. The latter is typically implemented as complex software tools, such as reservoir simulators.

We have built a multi-level expert system called ELAS (Expert Log Analysis System) for well-log analysis. Well logs are the various electromagnetic, sonic, and nuclear signals obtained from instruments placed down-hole in a well which characterize the properties of the rock and fluid formations below the surface. These signals are usually recorded and digitized on a per-foot basis for the entire depth of a well; they are displayed as continuous curves as in Figure 5.6. In this figure, several logs are displayed at depths between 11,200 and 11,300 feet. Each of the curves represents a specific measurent (whose identity and scale are not displayed in the figure). The appropriate logs are reviewed, processed and correlated, and then interpreted by the log analyst. From a practical applications point of view, well-log interpretation represents an important problem, since it permits an assessment of the likely presence of hydrocarbons and potential yields of the well

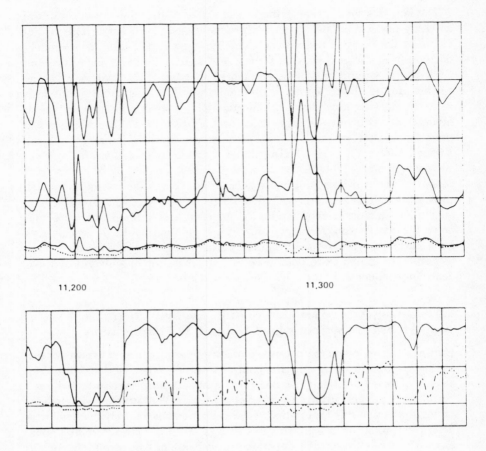

11,200   11,300

Figure 5.6:   Set of well logs before interpretation

during exploration and production. From the perspective of expert systems research, this application increases our understanding of representation, communication and control processes in multi-level systems. And, from the more general software engineering point-of-view, this system illustrates how one might exploit existing software systems more fully by building a coordinating and advisory system that makes these programs and their results easier to interpret by a wider variety of expert and non-expert users alike.

In many problem areas, it is not unusual to find that valuable software has already been developed to aid the expert in data analysis, the design of experiments, and the interpretation of results. These programs are often quite complex packages, developed over several years and enhanced through extensive user experience. In designing an expert system, it is only natural that one should want to take advantage of such software. One of the first efforts in modeling expert advice on the use of a complex program was the SACON project, which developed an advisory model for a structural analysis program. However, in that application, there was no.interaction between the two programs: SACON was run before the structural analysis program, giving advice on its prospective use. In order to develop an expert system to its fullest potential, interaction is needed between the advising program and the application programs. In a sophisticated system, the interpretive program should be fully integrated with the application programs so that they communicate their results to one another, and the advice changes dynamically as the model tracks the user interaction. Furthermore, the system must have the ability to automatically take a recommended action if the the user agrees. In effect, we will have a program that not only gives advice, but also can accept the advice and act on it.

5.4.1 COMPARISON WITH A CLASSIFICATION MODEL

An expert model which interacts with a complex program can be thought of as an extension of the classification model. Regardless of the underlying knowledge representation, these models are typically implemented as programs which ask questions of the user until enough evidence has been accumulated for the model to offer an interpretation. In situations where the human experts follow agreed-upon procedures for eliciting evidence, the questions are often highly structured, and one can expect that the results will be reported in a systematic and ordered fashion. Although some of the results may be gathered by external sources (such as instruments), they are typically filtered through the user who enters them into the program. In some situations, such as when most of the evidence is taken directly from instrument data or a data base, external programs feed evidence directly into the reasoning model.

In a model that is completely integrated with a set of complex application programs, so that they appear to the user as a single program, important additional functional characteristics are needed that go beyond those found in the typical classification models:

- The advisory model must be able to receive data or evidence automatically from the application programs, in addition to that reported by the user.
- The logic of the model must be prepared to interpret evidence in real-time.
- The model must be able to not only suggest advice on the interpretation of evidence, but must also monitor how the user reacts to this advice in

his subsequent choice of methods of analysis, and then provide new advice (to fit the dynamically changing situation) if the user requests it.

Yet, despite the above differences we can borrow an important analogy from the more traditional systems: the set of applications programs can be made to communicate with the advice model primarily through the evidence, as long as we are willing to broaden our definition of what we consider to be evidence. We will now take evidence to include not only facts about the particular domain problem, but also the status of the actions taken by the user while interacting with the applications programs, and the results of key calculations performed by these programs as the result of the user's actions. In an advisory model, we must represent classes and sequences of user actions, their consistency relationships, and expected effects. To be both effective yet flexible to the nuances of expert problem-solving, the advisory system must monitor and interpret the actions of the user in a type of "background mode." The user must have a considerable degree of freedom in the choice of methods and direction of analysis, yet the system must be able to detect choices which lead to poor decisions or introduce inconsistencies in the analysis or interpretation. If the analysis is proceeding smoothly, as might be expected with an experienced user, the model should be available in an advisory and summarizing capacity, without interfering with the user's control over the problem-solving flow.

The logic for the interpretive analysis can be stated in the form of production rules as in the usual classification models. However, the designer of the model must take into account the newly enlarged scope of types and the dynamically varying nature of the evidence that must be handled. While in a traditional consultation system we expect evidence to remain relatively stable during a single consultation session, here, because the control and monitoring of the separate applications programs must be carried out in real time, evidence will change values very frequently in the course of a session. For example, an observation such as *Task A not performed* will be changed once Task A is done; or a numerical result may be received from one program, and then be changed as the consequence of further analysis and processing by another program. This type of situation resembles a consultation where someone is continually modifying or updating the evidence, and sequentially asking for an interpretation.

Another important difference with the usual classification model is that the controlling model has the opportunity to go beyond just giving advice; it can provide the means of accepting the advice and taking the recommended action in real time. This step may prove somewhat more difficult for expert system builders than might be initially expected, since the model and the set of application programs must be unified in what appears to the user as a single system. In addition, a single task of the expert may require the compilation of many steps through one or several of the programs or even combining or adding steps that are not directly available in the application programs. Although the application programs may not have to be completely recoded, new code may be needed to integrate steps that reflect the expert's particular approach to analyzing a problem. In our experience, we have found that this can be done successfully while building an overall expert system.

5.4.2   ELAS: AN EXPERT SYSTEM FOR WELL-LOG ANALYSIS

ELAS (Expert Log-Analysis System) is a demonstration of how the knowledge and reasoning methods of an expert log analyst can be combined with Amoco's large-scale interactive program for well-log data analysis and display. Working with the expert analyst, we have observed the sequence of steps involved in solving certain typical interpretation problems and encoded them as a set of advice rules using the EXPERT formalism. Thus, more specific objectives in developing ELAS include:

- formalizing methods of expert well-log analysis into a representation that can be easily used and understood by others
- providing interpretations based on the data, the actions, and expectations of the user and the program.

ELAS uses a two-terminal configuration consisting of a graphics terminal (with a joystick), a video display terminal, and a single alphanumeric keyboard. The system allows the user to interactively perform experiments in the analysis of logs. Advice is generated based on the results of previous experiments, and a running summary is kept of the actions already taken. The advice system is integrated with Amoco's well-logging software, while the interpretations are based on the EXPERT production rule scheme.

One of the contributions to a specific domain that comes from building an expert system is that it helps to structure and organize practical problem-solving knowledge in the domain. When many of the important methods of interpretation are informally specified and are personal to the experts in the domain, the expert systems formalization can be particularly valuable. By encouraging or even forcing a formal structure on the domain knowledge and incorporating it into an expert system, the knowledge becomes testable, reproducible and more widely reportable to others. While many general software programs require expertise not only in the application domain but also in the use of the program, our objective in developing ELAS has been to produce a new, powerful yet easy to use experimental tool for the well-log analyst.

To make interaction easy for users, the front-end of the ELAS system has as its top level a *Master Panel*, which holds a snapshot of the current status of the analysis of an already selected well (Figure 5.7). Most user-program communication is controlled through this master panel. It is displayed on the

HIGHLAND PARK NO. 2					
SURFACE TEMP.	50.	TEMP. GRADIENT	0.015		
R_{MF}	2.000	R_{MFT}	65.		
R_M	2.000	R_{MT}	65.		
		BITSIZE	8.50		

ZONE	1	2	3	4	5
BACKGROUND					
TOP	1060.	1850.	2643.	3400.	3901.
BOTTOM	1340.	2300.	3399.	3900.	4500.
LITHOLOGY	CLEAN SAND	CLEAN SAND	CLEAN SAND	CLEAN SAND	CLEAN SAND
FLUID	GAS WATER	GAS WATER	GAS WATER	GAS WATER	GAS WATER
DISCRIMINATORS					
BAD HOLE					
MAX. POROSITY	40.	40.	40.	40.	40.
LITHOLOGIC					
SP GREATER THAN	-10.	-9.	-13.	-7.	-15.
GR GREATER THAN	39.	47.			
POROSITY					
METHOD/INTERPRETATION	NP-DP XPLT	DP	VALESKE	DP	DP
R_{HOG}		VARIABLE	2.65	VARIABLE	VARIABLE
R_{HOF}		1.0	1.0	1.0	1.0
R_{HOGAS}			0.23		
Z_{DN}			0.30		
R_W INITIAL					0.2000
R_W QUALITY	UNKNOWN	UNKNOWN	UNKNOWN	UNKNOWN	FAIR
R_{WT}	100.	100.	100.	100.	100.
R_W USED			0.5599	0.9487	0.2000
R_W METHOD	INITIAL	INITIAL	PICKETT	1/SW HSTGM	INITIAL
S_W EQUATION	ARCHIE'S	ARCHIE'S	ARCHIE'S	ARCHIE'S	ARCHIE'S
N INITIAL	2.0	2.0	2.0	2.0	2.0
M	1.40	1.80	1.80	1.40	1.80

Figure 5.7:   Master panel

graphics screen, and includes a concise set of key parameters and tasks that are crucial in a well-log analysis session. Here, a parameter may be a constant, a log (represented as a vector of digitized values for each foot of depth in the well), or an expected characteristic of the well (e.g. the presence of gas in some zone). The value of the parameter will influence the results of the subsequent analysis. Whereas the original software was an interactive system that required specification of many detailed operations, our task has been to develop a much higher level system that compiles many of these smaller steps. At the initial user level there is a superficial similarity to the popular personal computer spreadsheet programs. In the simpler microcomputer environment of the spreadsheet calculators, we see a program that presents information in a concise format and allows the user to vary a parameter and then watch all dependent results change. In our case, we are faced with a much more complicated computational task, but we too try to show the propagation of effects that follow from the user's change of a parameter value or choice of analysis method within as short a time as possible, ranging from almost instantaneous to many seconds. This is done by updating the master panel at the top level, from which the user will be led to more detailed panels or displays for the specific methods. Changing a parameter may imply quite a large number of computational steps and not all information can be described in a simple tabular format.

In ELAS the user can direct both the mathematical analysis and the interpretive analysis by changing parameters or invoking tasks through the master panel. As just stated, the outcomes of mathematical analyses that follow are then reported back to the user through this same panel. The expert system keeps updating its interpretive analysis after every change in the evidence so that it always reflects the current status of the panel. Changes are made through either user action or updates in the mathematical analysis. The user has the freedom to carry out an entire well-log analysis sequence without ever asking for advice from the system, or he can request advice for any stage where he feels the need or curiosity for it.

The example in Figure 5.7 shows the master panel for a sample test well. The five columns correspond to five different geological zones (by depth) which have been chosen for analysis. The rows correspond to the parameters for the zones. Initial values for some of the parameters must be supplied by the user. Many of them, however, may be obtained through subsequent analysis. Some parameters may stand for specific tasks that the user might want to invoke to help in the analysis.

Consider the case where the user wants to examine (on his own or following the advice of the model) the type of geology and how it may affect the calculation of rock porosity in a zone. The appropriate selection is made from a menu which appears on the terminal, as illustrated in Figure 5.8 where PF2 (programmable function key 2) is used.

ZONE 1

SELECT A POROSITY COMPUTATION METHOD:

PF1	DENSITY POROSITY
PF2	NEUTRON-DENSITY CROSSPLOT
PF3	HOUSTON GAS CORRECTION FOR SAND/SHALE
PF4	VALESKE METHOD

$$\vdots$$

PF11	RETURN
PF12	HELP

Figure 5.8: Menu for methods of porosity computation

The system invokes the task, displaying first a crossplot of the zone depth points with the various geological choices made available for further analysis (Figure 5.9). Once the user has selected various parameters the program performs various related calculations. The program then returns to the main panel, which now reflects any updates due to this task invocation.

The system also maintains consistency between dependent tasks, which is necessary whenever a significant parameter changes values. This can involve computations ranging from almost negligible to very formidable. For example, one of the most frequently used formulas in log analysis is:

$$S_w = (F \cdot R_w / R_t)^{.5} \tag{5.1}$$

where the variables are quantities that can be changed through the master panel. Even a minor change in one of the variables in this equation involves a recomputation for all the points in the log, which are usually in the thousands. This will immediately cause reinterpretation and revision of previous conclusions and recommendations.

In summary, the formalization of the methods in a domain may very well require a new organization and presentation of the original software, and may not be quite as simple as feeding numbers and arguments back and forth between the original software and the model. A clean interface is of course an important ingredient in the ultimate success of the program. In our case, the system will lead the user into many specialized routines associated with a particular task. Often, the setting of a key parameter may require a separate analysis in itself.

The well-log analysis model considers three types of advice which can be described in a classification model framework.

1. interpretation of existing evidence and past actions
2. advice on future actions
3. consistency checking.

In a system such as ELAS, there are many possible reasoning paths that the user may take, most of which are under user control. Thus depending on the user's prior actions, the model may give radically different advice even for the same initial data. Furthermore, depending on its interpretation of the status of analysis and user actions up to a given stage, the model will make recommendations to try out various preferred methods for subsequent analysis. The user also has the choice to enter certain *a priori* information about the problem, such as whether one ought to expect gas in the well. In this case, if we were told not to expect gas, but gas is indicated by an analysis of some of the logs, we can proceed to get clues as to whether the method of analysis might be at fault, whether the logs are noisy or otherwise inaccurate, whether some underlying assumption is unjustified, etc. This is an example of the kind of consistency checking that involves both the log data and the inter-

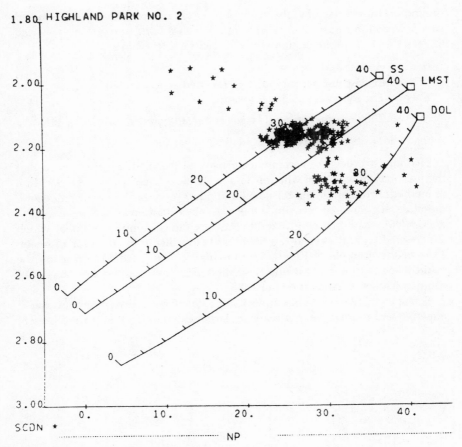

Figure 5.9:   Neutron density crossplot

play between the various methods of analysis. While actual advice depends
on the types of analyses that are specific to a domain, we have found that they
are representable in a classification model framework.

5.4.3   COMMUNICATION BETWEEN PROGRAMS

Because we are starting with existing software, we initially see two separate
programs, the original Amoco software (INLAN) and the interpretation program (EXPERT). To integrate the two, we need a means of communication,
which involves automatically filling in the arguments of some of the evidence
necessary to correctly invoke the production rules. Figure 5.10 is a simple
illustration of such a production rule, where the finding of gas can be com-
municated once certain tasks have been performed in the well-log analysis

program. This requires that the original program record such information and pass it on to the interpretation program. Because both programs are written in FORTRAN, communication is relatively straightforward.

If: gas is found and
          the current porosity log has not been gas corrected
Then: The following advice is given:

The porosity logs may be gas corrected by methods a, b.

Figure 5.10:   Example of a rule for advising on methods

Secondly, we need to communicate back to the well-log program so that the user may now choose whether to accept the advice or not. The selection of methods through the master panel and other screens is menu oriented and, therefore, we can use a dynamic scheme to place an item on the menu. A production rule may then be invoked to indicate the circumstances under which the method is displayed on the menu and is available for the user to select. Thus in the example of Figure 5.10 methods *a* and *b* would appear on the menu when the production rule is satisfied. Figure 5.11 gives an overview of communication channels used in ELAS.

A major challenge of this project is to develop an expert system that is effective and productive in a realistic, large-scale application. In the present

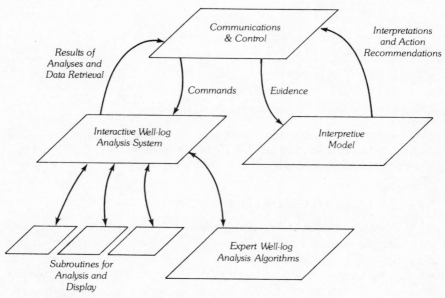

Figure 5.11:   Overview of the multi level expert system ELAS

project there is a strong incentive to encode expert knowledge and make this knowledge available to others in the field because of the scarcity and cost of expert interpreters, and for the economic benefits of more accurate interpretation. Another long-term research goal is the development of more general schemes to interface an interpretive program with existing applications software.

5.4.4   INTERFACING THE ADVICE WITH THE APPLICATIONS PROGRAM

While excellent tools can be developed to aid in the design of an expert system, the completed application system will most likely require some specialized programming that will be determined by the problem domain. For example, it is clearly useful to have a facility to summarize those findings which are positive for a given case. Such summaries, as illustrated in Figure 5.12, are necessary for developmental design and implementation work. Frequently, however, knowledge of the domain will allow for the presentation of such information in a more suitable form. One can design summaries that list only the most important findings and which organize these findings by topic. The master panel of Figure 5.7 is an example of a transformed summary that has been developed for the well-logging domain. This summary includes only a small subset of all findings for a case. These are the key findings necessary to describe the methods and parameters for the computation. Many other findings are not presented to the user, except when appropriate as part of the interpretive analysis. The complete list of findings is too detailed for the end user, but such findings are important in debugging and testing the system. As with all programs, an appealing user interface is of paramount importance. The completed user/program interface will often depend on the type of application and the hardware to be used in program. For example, for ELAS we are using two interacting terminal displays: a graphics terminal and a regular video display terminal. For this problem, the use of programmable function keys has proven extremely useful. While the user interface is very important in the acceptance of the system, the internal representations (such as a set of production rules) can be quite similar for most classification models. The character of the findings may, however, be somewhat different. They often may be the result of tracking actions that the user takes, or they may be characteristics of the methods that are invoked by the user. For example, once a particular procedure is invoked by the user, the system may go on to record various characteristics of the results of the procedure, such as performing a statistical test like the goodness of fit to the normal distribution. This characteristic can then be recorded as a finding to be used in a production rule.

SUMMARY

Tasks performed:
  Log display
  Neutron-density crossplot
  Lithological discriminators
  Sw computation
  1/Sw histogram plot

•
•
•

Mode of 1/Sw histogram: 1

1/Sw histogram characteristics:
  Distribution around 1 and normal (bell) shaped
  between 0 to 2.
  Many values found above 2

•
•
•

INTERPRETIVE ANALYSIS

Rw seems reasonable for the Sw computation.
The distribution of the 1/Sw histogram tends to
support the hypothesis that significant hydrocarbons
are present.

•
•
•

Figure 5.12:   Summary and advice at end of session

Figures 5.10 and 5.13 illustrate the types of advice that can be given by a
system such as ELAS. In Figure 5.10 the system may give a simple recommendation for selection of methods. The actual procedures to perform the
task are compiled procedures. This example illustrates several characteristics
of a model which controls another program. The findings are set automatically as the user performs various tasks. They are not questions answered by
the user. As a session continues, the rule at any given time may or may not be
satisfied. This would be seen by the user as a menu item which may or may
not appear on the menu for actions to be taken. By selecting this menu item,
the action which corresponds to recommended advice may be taken. In the
second example of Figure 5.13, we see a somewhat more complicated analysis

for the interpretation of results. The components of the production rule can readily be viewed much as any other findings. Clearly, these findings require a fairly complex procedural analysis in the form of separately coded routines which supply the necessary information. Thus we see that the writing of the rule is much simpler than the task of obtaining the necessary information.

If: There are 30 or more points in the water zone and
          there is a good fit for $1/S_w$ with
          either the log normal or sqrt normal distribution and
          the quality of the initial R_w is good
Then: The following interpretation is given:

The water zone is consistent with good R_w values.

Figure 5.13:   Example of a rule for interpretation

5.5   Extensions to the Classification Model

The design philosophy for expert systems we have emphasized throughout the book is one of choosing a relatively stable computer representation, seeing which problems can be successfully solved using it, and then examining where the systems built with this representation fall short in specific real applications. We then try to augment the original representation to allow for the solution of a larger class of problems. Clearly, if there are major deficiencies in the original representation, or if the original representation is inappropriate for the the new class of problems, then this approach will be unsuccessful. We have seen, however, that with the relatively simple, yet well-understood classification model, a large class of problems can be successfully handled. It is quite natural, then, to try to extend this representation to handle problems that will not fit into the pure classification mold; but with slight changes in representation and perspective, they can be viewed as augmented classification problems.

An analysis of several application systems, including ELAS, DASD, and R1, suggests an extension to the pure classification models which were described in Chapter 4. If we look at the hypotheses as stated in the classification model, we see statements which can be assigned a confidence measure indicating a degree of belief in their presence or absence. The confidence factor is the sole dynamic factor in the evaluation of the hypothesis. For example, in the car repair model, we had a hypothesis indicating that the battery was dead. During a session the confidence in this hypothesis may be assigned a numerical weighting from -1 to 1. This will be satisfactory for a great number of classification problems. One could take, however, a broader view of the definition of a hypothesis which would include recommendations for taking some *action*, or even a number of actions (if they are under the control of the program). The simple car repair model had a relatively elementary function which could compute miles-per-gallon for gas consumption. Such simple functions are easily anticipated, but more complicated procedural knowledge can be more difficult to handle. In order to fit within the general classification model, the great majority of hypotheses usually would have to be classificatory in nature. However, some production rules might be of the form:

$$H(Class,.1:1)\&\ldots \rightarrow H(Action)$$

where instead of setting a hypothesis with a specific confidence measure, some action is taken. In the case of ELAS, for example, the action might be that a statistical routine is invoked. The results of this routine (and in general the results of these *actions*) would be reported as typical findings used in production rules, such as the mean or standard deviation of a newly generated array of log points. In the case of the DASD system, the action is usually relatively simple, such as getting a specific byte from a machine register or checking whether a specific bit is set. The premise of the rule would indicate which byte is appropriate by taking into account various factors such as the specific model of storage device. In the case of the R1 system, the classification model is stretched the most, since many of the production rules are actions rather than pure classifications. Even these, however, can be viewed as classificatory if we extend our definition of classification. In R1 the ultimate goal is to assemble a VAX configuration. Many of the production rules result in actions (classifications) of attaching a specific device to the current dynamic configuration. The system starts out with a list of components and dynamically builds the configuration. At any instance in this assembly process, a decision must be made on the next component to add to the current partially assembled configuration. Thus the production rules indicate which action should be taken and the list of potential actions is well-circumscribed, as in any classification model. However, an extra factor is the current diagram of how everything fits together. Because many of the findings in the model are dynamically derived from the partial configuration at any given moment, the current partial configuration is consulted by the production rules before the next component is added, and the new incomplete configuration diagram is the basis for re-invoking the production rules afresh. In general, knowledge of the application and the type of system design will dictate whether previously satisfied production rules need be re-evaluated after new actions have been taken, producing new findings.

The classification model is an excellent starting point for building many expert systems. Even when it does not fully cover a specific application, there may well be strong classificatory aspects to the problem. In those cases the designer of the system must show flexibility in the original model specifications, so that the classification model can be used for somewhat different problems.

Now that we know how to build an expert model, we will next examine how we may test, evaluate, and improve the performance of an expert system.

5.6 Bibliographical and Historical Remarks

The Serum Protein Diagnostic Program was first reported in Weiss, Kulikowski, and Galen (1981). It is now available at dozens of clinical laboratory sites. The AI/Rheum system, which is the name of the rheumatology consultant, is a large collaborative project between research groups at Rutgers University and the University of Missouri—Columbia. It was first described in Lindberg, Sharp, Kingsland, Weiss, Hayes, Ueno, and Hazelwood (1980). The notion of frames or prototypes was first introduced to a medical consultation system, PIP, by Pauker, Gorry, Kassirer, and Schwartz (1976) and was also used extensively in the CENTAUR system (Aikins 1983). Some interesting thoughts on the relationship between deep and surface models of reasoning are presented by Hart (1982). One of the earliest systems for advising on the use of another program was SACON, which used the EMYCIN framework, and was reported in Bennett and Engelmore (1979). Instead of a stand-alone program as SACON, the advisory program for the ELAS system is fully integrated with the programs for well-log analysis. ELAS was first reported in Weiss, Kulikowski, Apte, Uschold, Patchett, Brigham, and Spitzer (1982). Several other systems have been developed for signal-processing applications, including VM (Fagan 1978) and HASP/SIAP (Nii, Feigenbaum, Anton, and Rockmore 1982), and the dipmeter advisor which is a highly specialized well-logging program developed by Schlumberger (Davis, Austin, Carlbom, Frawley, Pruchnik, Sneiderman, and Gilreath 1981).

6

Testing and Evaluating an Expert System

6.1 Introduction

During the course of building an expert system many changes are likely to be made to the original design. Both implementors and experts will typically change their ideas on many of the design details, and sometimes even on what the system is meant to accomplish before they are satisfied with a "final" design. Yet, as the changes are being made, one would like to ensure that, to the greatest extent possible, these changes will not damage any of the successfully working components designed and implemented earlier. This is particularly important when the system becomes more mature and covers a number of difficult topics reasonably well. If the design reaches a plateau, where it seems to be performing quite well by informal observation, how do we then go about evaluating the performance more formally? How do we combine performance measures for one set of subproblems with those for another? What measures of overall performance should we use? In short, how do we support our claims that the system is performing at an expert level for a well-defined, circumscribed problem?

There are two approaches which illustrate the two extremes in evaluating expert system performance. One we will call the *anecdotal approach*, the other the *empirical approach*. With the anecdotal approach the model designers describe their good experiences, or those situations where the program performed well. They describe the situation, perhaps even recreating a session with the program that duplicates the original performance. In those situations where the program performs poorly, attempts are made to correct the program. As new problem situations arise, new information is incorporated into the model, but if the number of problems covered by the model is large (as it is almost always bound to be in expert systems), the designers typically only maintain a small amount of short-term memory to check on possible changes induced in past problem cases by the current changes in the model. One then can only hope that the new changes will not create any significant errors in cases that were handled well previously. Like human intelligence, we expect that the performance of an expert system will improve with experience over time, but unlike people, current expert systems lack powerful *learning* capabilities, and we are sometimes surprised that the system misses cases that should be easily handled and which were correctly handled previously.

The second approach places its emphasis on the empirical evaluation of performance over many problem cases stored in a data base. Some kind of rigorous testing procedure must be specified to compare the model-produced interpretations to independently obtained, conclusive interpretations for the same problem cases. Traditionally, the testing is carried out once the system is relatively well developed. However, empirical testing can be of great value even at the early stages of model development. For this, a representative sample of cases must be available in the data base, so that when changes are made to the model, the performance of the system can be tested by examining the changes in performance for the cases in the data base. We shall see that there are several types of tools that can be developed to use such information in helping design an expert model.

While an empirical approach to testing may be clearly superior to an anecdotal one, implementing the methods and obtaining the representative cases for the data base often comes up against severe practical obstacles. In some domains, such as medicine, it may be possible to gather large numbers of cases for the relatively common diseases, but rare diseases always present a problem in getting enough representative coverage for validation. In other domains, such as geological exploration, the cost of obtaining sample cases may be very high, and only a few of any type of ore formation may be readily available. In PROSPECTOR, for instance, a special sensitivity analysis procedure was developed to compare the results of the model-produced interpretation to the conclusive results of exploration. This was done partly to overcome the limitation on the number of cases available for testing.

There are also more fundamental difficulties with the purely empirical approach. For the analysis to be accurate and useful, one must have firm end points; that is, the correct conclusion for each of our data base cases must be known. Then one can judge performance on an absolute scale: that of the proportion of correct-to-incorrect decisions. In some models, though, this may not be enough. A more detailed analysis might be valuable, breaking the decisions into types and analyzing the results by category. In medicine, for example, the percentages of correctly diagnosed cases for each disease, together with the percentages of cases misdiagnosed by each possible category of mistake, are essential in assessing the performance of an expert system.

All of these evaluations ideally reduce to a binary decision: correct or incorrect. Unfortunately, not all problems can be easily categorized this way. The classification model does lend itself to this type of analysis. Even within this framework, however, there are many ways of describing a model without clearly defined endpoints. For example, if we use a model that produces narrative as interpretational output from the concatenation of several statements that are applicable conclusions for the case, we are faced with an overall conclusion that may be quite satisfactory to the user, but which is hard to break into firm endpoints. In such cases it is difficult to ask the experts to independently evaluate the cases and to then compare results, since their unconstrained narrative statements may prove to be very different from those of the system. Even if the expert were asked to choose from the system's list of conclusions, there will be problems in combining these statements into an overall interpretation. This is what we experienced in the case of the Serum Protein Diagnostic Program, where the scope of statements is quite broad and where they are not mutually exclusive, so that it is difficult for the experts to reach exact matches. In these situations, what is commonly done is to show the expert the system's results and ask whether he agrees with the conclusions. Although it may introduce biases which are avoided in the *blind* method, this form of evaluation is one that is most frequently used for practical reasons and it was the one we used for evaluating the Serum Protein Diagnostic Program. Several independent experts may be employed for the evaluation, but they all judge the system's results, rather than providing an independently determined interpretation for comparison.

6.2   Tools for Model Evaluation

In this section we describe various tools which are relatively easy to build and which help the model designer evaluate the performance of a model. These tools can aid the model designer both in an overall evaluation using a large sample of cases, or in the more detailed analysis of what went wrong in a specific case. We will then look at the decision-table model described in the preceding chapter, and see how a data base of cases can be used to not only evaluate the application model, but also to provide a wealth of information for guiding the model designer in improving the performance of a model. In this particular type of analysis, firm endpoints are needed in the analysis.

6.2.1   CHECKING THE CONSISTENCY OF RESULTS FROM MODEL CHANGES

An important tool in designing and evaluating expert models is a consistency checking procedure which lists all differences between the conclusions for the stored cases before and after a model has been modified. During the course of a consultation, a case can be stored in a data base along with the weights assigned by the program indicating the likelihood of each hypothesis. Modifications of the knowledge base usually result in new measures of belief being computed for the conclusions applicable to a case. A common modification would be to change a production rule so as to correct the conclusion of the case. It is important to consider the effect of this modification on the conclusions of all cases stored in a data base. It may not be necessary to consider small changes in weight assignments, but only those which exceed a specified threshold, signifying a change in the final qualitative interpretation of the case.

During the course of design, the findings and hypotheses are often changed quite frequently. In order not to lose the contribution of stored cases after modifications to the list of findings or hypotheses, the program should automatically reformat the data files to maintain consistency with previously stored cases. This is relatively easy to do when we use consistent sets of mnemonics for the findings and hypotheses in both the old and the revised model. For example, using mnemonics for each hypothesis, Figure 6.1 describes the effect of a change on the performance of a model for a particular case. The program output shows the changes for each hypothesis (as indicated by the mnemonics on the left). In Figure 6.1 a test case has had the confidence weight assignments for all the listed hypotheses changed as the result of a modification in the reasoning model. Other hypotheses whose weight assignments were not changed as a consequence of the model revision are not listed. After examining the effects on all the cases in his data base, the model designer can make further changes in reasoning rules or data elements if the results are not satisfactory. In this way the designer can carry out a quick interactive sensitivity analysis of the model. The process of model construction can be speeded up considerably by this procedure.

6.2.2 SEARCHING THE DATA BASE FOR PATTERNS

A search program which allows an interactive search for patterns of findings and hypotheses over the data base of cases can be extremely valuable in

[PROCESSING CASE 5]
Cadillac

CHOKE   .8 → .9
FILT   .6 → .2

[CASE 5 PROCESSED]

Figure 6.1:   Checking consistency of rule changes

analyzing the performance of the model. We assume that not only data, but also hypotheses with their computed confidence weights, are stored in the data base. A search pattern can then specify different decision thresholds on the hypothesis weights as well as different patterns of findings. Techniques for searching a data base for specific patterns can provide an invaluable tool, both in formulating decision rules and in checking the consistency of subjectively constructed decision rules.

The following abstracted session from the auto repair model illustrates how one might build a data base system to retrieve statistics for all cases with a specific pattern.

We are looking for all cases where the program concluded that the choke was stuck (with confidence $> .5$) and the temperature was less than 40 degrees.

Enter SEARCH CODE Mnemonics:
SEARCH: **CHOKE$>$.5 & TEMP $<$ 40**

The program responds with the full English version of the search pattern so that the user can check for correctness, and then indicates that two cases, 3 and 5, satisfy the pattern conditions.

Choke Stuck is greater than 0.5
And
Outdoor Temperature (degrees F): is less than 40

Cases satisfying conditions: 3   5
            Cases: 2/   5   (40.00%)

The features of these two cases are summarized next. The summary provides a statistical comparison of the results of all findings for those cases that satisfied the conditions (marked with a +) versus those which failed the conditions (marked with a −). During a session, not all questions are asked, so not all findings are reported for each case. An "" indicates statistical significance. MNE is the test mnemonic, CAS is the number of cases. This example is illustrative only; the sample is too small for any valid statistical results.*

= = =DATA BASE SUMMARY= = =
=FREQUENCY DISTRIBUTION=

TEST MNE	FREQ+ CAS	PCT	FREQ− CAS	PCT	CHI SQUARE	TEST ENGLISH
						Type of Problem:
FCWS	2	100%	3	100%	0.000	Car Won't Start
						Odor of Gasoline in Car:
NGAS	0	0%	1	50%	5.333*	None
MGAS	2	100%	1	50%	0.000	Normal
						Simple Checks:
DIM	0	0%	2	67%	5.868*	Headlights Are Dim
						Starter Data:
SCRNK	0	0%	1	50%	5.333*	Slow Cranking
OCRNK	2	100%	1	50%	0.000	Normal Cranking
EGAS	0	0%	1	50%	5.333*	Gas Gauge Reads EMPTY

=NUMERICAL DISTRIBUTION=

TEST MNE	CAS+ CAS−	MEAN+ MEAN−	SD+ SD−	T-TEST SIG	TEST ENGLISH
TEMP	2	17.500	10.600	−7.975	Outdoor Temperature (degrees F):
	2	78.500	2.121		
CWS	2	0.600	0.000	0.189	Car Won't Start
	3	0.534	0.472		
FUEL	2	0.600	0.000	0.774	Fuel System Problems
	3	0.300	0.519		
FLOOD	2	0.001	0.000	−0.775	Car Flooded
	3	0.001	0.000		
CHOKE	2	0.644	0.000	999.	Choke Stuck
	3	0.001	0.000	**	

•
•
•

Would you like a SUBGROUP search? *y

This command allows us to search further within the group of two cases that satisfied the initial search pattern. We next enter the specification of the search pattern for the subgroup, which searches for temperatures below 20. One of the cases, number 5, does satisfy this condition.

Enter Subgroup 1 SEARCH CODES:

SUBGROUP: TEMP$<$20

    Choke Stuck is greater than 0.5
And
    Outdoor Temperature (degrees F): is less than 40
And   (Subgroup 1)
    Outdoor Temperature (degrees F): is less than 20

Cases satisfying conditions: 5

    % of Subgroup: 1/2 (50.00%)
    % of all cases: 1/5 (20.00%)

•
•
•

It is particularly valuable to be able to treat the hypotheses of the model and their assigned confidence measures no differently than the recorded measurements for the cases. Thus one might search the data base for those cases where hypothesis X $>$ .5, i.e. search for those cases where the system concluded hypothesis X with a a confidence greater than .5.

6.2.3   MATCHING THE COMPUTER'S AND THE EXPERT'S CONCLUSIONS

A very simple but valuable tool can be applied when one has the expert's conclusion stored for each case in a data base of cases. Such information is not always available, but when it is, one can obtain a definitive evaluation by matching the expert's conclusions to those of the model. The cases for which the system does not match the expert can then be identified. This is particularly useful after a change has been made to the model. In Figure 6.2, those cases for which a mismatch occurred are listed; the model's first and second choices (by confidence level ranking) are both listed together with their computed confidence measures. One can then observe the relative closeness of the strength of belief in the two highest ranked conclusions produced by the model.

Thus we obtain a rapid synopsis of results for all cases in the data base which can serve to suggest a more detailed investigation of specific discrepancies.

Case	Physician	Model's 1st choice		Model's 2nd choice	
1	MCTD	MCTD	0.550	SLE	0.500
2	RA	RA	0.850	SLE	0.500
3	PSS	PSS	0.700	MCTD	0.600
4*	MCTD	SLE	0.700	MCTD	0.550
5	SLE	SLE	0.900	RA	0.000

•
•
•

Figure 6.2: The computer versus the expert's conclusions

The previous discussion illustrates some of the relatively simple tools which can be used to evaluate a model's performance on a data base of cases. These tools supplement any methods we use in the analysis of a single case. In the single-case situation, we concentrate on examining those rules which were satisfied and which caused a specific conclusion to be reached or not reached. In the next section, we will look at a more complex set of tools that suggest possible changes to a model based on an analysis of the kinds of matchings described here.

6.3   Learning from Case Experience

A primary direction for research and development of expert systems has been the acquisition of knowledge from the expert. Less attention has been given to finding effective methods for validating a system's knowledge base and evaluating its performance. A set of generalized knowledge engineering tools can facilitate both the building and testing of an expert system. We have seen several examples of generalized knowledge engineering tools for the classification model. These systems provide the expert-system builder with a prespecified control strategy, a formalism for encoding expert knowledge, explanatory tools for tracing the execution of rules during a consultation session, and a data base system in which cases can be stored for empirical testing. Work on empirical testing of expert systems has been reported in the development of the PROSPECTOR consultation model for mineral exploration. The PROSPECTOR scheme uses sensitivity analysis to determine the effect on the model's conclusions that result from making changes to the levels of certainty associated with the input data. For PROSPECTOR, empirical testing is based on matching the expert's conclusion to the program's conclusion and the intermediate conclusions reached by the model.

The TEIRESIAS system was one of the first systems developed to assist the model designer in explaining an expert system's decisions. During a consultation session with the MYCIN system, TEIRESIAS assisted the user in composing new rules that would correct erroneous conclusions. TEIRESIAS generated its advice about the contents of a new rule by using "meta-rules" containing knowledge about reasonable changes to the rules in the MYCIN knowledge base. It did not, however, directly determine the impact of changes to the knowledge base on other cases previously processed by the consultation program.

An experimental approach to performance analysis that we will describe integrates performance information into the design process of an expert system. A system called SEEK has been developed by Politakis to generate advice in the form of suggestions for possible experiments that will either generalize or specialize the rules in an expert model. Performance measures derived from stored cases with known conclusions are used to interactively guide the expert in refining the rules of a model. The decision-table model described in section 5.3 is the representation used for capturing the expert knowledge and used for the empirical analysis. As was illustrated in Figures 5.4 and 5.5, these models are readily translated into production rules.

6.4   Stages of Model Development

To perform an empirical analysis, SEEK must have a tabular model defined for each final diagnosis. Also required is a data base of cases, including the correct final diagnosis assigned to each case. The design of a model and the analysis of performance can be broken down into the following steps:

- initial design of the application model
- data entry: cases and expert's conclusions
- performance analysis of the model
- generation of model refinement experiments
- review of the effect of model changes on the case conclusions.

6.5   Design of the Model

Because SEEK operates on a highly structured class of models, a specialized text editor is used to specify an initial design of the model. For each newly identified final conclusion, the model designer can list major and minor observations and specify components of the rules that would reach that conclusion. When additions and updates are made, a table is stored and translated into a format used by SEEK. The translation of the table is completely compatible with the EXPERT system format. The SEEK translator translates an application model into an EXPERT model, assigning mnemonics to hypotheses and findings. These mnemonics will have keyword prefixes which will be used by SEEK in its analysis. The model described in section 5.3 is the model used by SEEK. A sample output of the consultation system was described in Figure 5.3.

6.6   Analyzing Model Performance

The process of improving the performance of an application model typically would consist of iterations through the following steps:

- obtain the performance of rules on the stored cases
- analyze the rules
- revise the rules.

In reviewing the performance of a model, the expert's conclusions are matched to the model's conclusions. The expert's conclusion is stored with each case, while the model's conclusion is taken as the one reached with the greatest certainty. Figure 6.3 gives an overview of the process of performance analysis. The SEEK system attempts to recommend potential refinements in the model. Even without such automated procedures, the empirical review of the performance of the rules should provide valuable information to the knowledge engineer.

6.6.1   SUMMARIZING MODEL PERFORMANCE

The starting point in any performance analysis is a measurement of performance of the model on a data base of stored cases. Figure 6.4 is an example of a summary of performance for a subset of a medical model. Performance is evaluated by matching the expert's conclusion to the model's conclusion in each case. The results are organized according to final conclusions and show the number of cases in which the model's conclusion matches the expert's conclusion. The column labeled False Positives shows the number of cases in which the indicated conclusion was reached by the model but did not match the stored expert's conclusion. In the computer listing of Figure 6.4, the summary of performance for mixed connective tissue disease (MCTD) indicates that 9 cases out of 33 were correctly diagnosed. Furthermore, there are no cases which were misdiagnosed by the model as being mixed connective tissue disease when they were really not. This shows that the rules for MCTD are not picking up the disease nearly as well as we would like! In contrast, the rules that conclude rheumatoid arthritis (RA) perform quite well for the stored cases of this disease. Yet, they are not perfect, because they are picking up 9 false positives over and above the RA cases. This makes them candidates

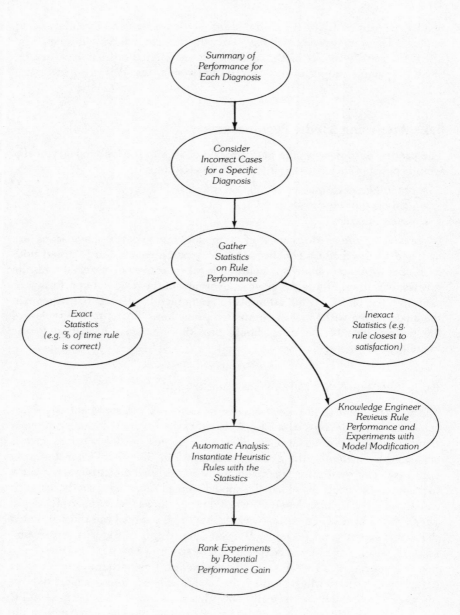

Figure 6.3: Overview of rule performance

			False Positives
Mixed Connective Tissue Disease	9/ 33	(27%)	0
Rheumatoid Arthritis	42/ 42	(100%)	9
Systemic Lupus Erythematosus	12/ 18	(67%)	4
Progressive Systemic Sclerosis	22/ 23	(96%)	5
Polymyositis	4/ 5	(80%)	1
Total	89/121	(74%)	

Figure 6.4:   Example of a summary of overall current performance

for specialization, or *strengthening* of their antecedent conditions so that they should filter out the non-RA cases more effectively.

In addition to the results shown in Figure 6.4, general performance results about a specific rule can be obtained showing the number of cases in which the rule was satisfied. An example of this is shown in Figure 6.5. This summary of a rule's performance includes the number of cases in which the rule was used successfully (i.e. in which it matched the expert's conclusions) and the number of cases in which the rule was used incorrectly (i.e. in which it does not match the expert's conclusions).

The overall performance measures for the data base of cases will generally suggest to the model designer a single conclusion for which system performance needs improvement. This will focus attention on a subset of rules in the model, particularly the subset of rules which are relevant to the misdiagnosed cases for the selected diagnosis. If we concentrate solely on gathering performance statistics for the rules, we will have a basis for experimentation

Rule 72:   2 or more Majors for RA
                2 or more Minors for RA
                No Exclusion for RA
                → Probable Rheumatoid arthritis

43 Cases: in which this rule was satisfied.
13 Cases: in which the greatest certainty in a conclusion was
                obtained by this rule and it matched the expert's
                conclusion.
7 Cases: in which the greatest certainty in a conclusion was
                obtained by this rule and it did not match the
                expert's conclusion.

Figure 6.5:   Example of summary of a rule's performance

with refining the rules. SEEK tries to go beyond the gathering of these statistics; it attempts to suggest specific experiments for rule refinement. Heuristic procedures are needed to select experiments from the many possibilities. For example, SEEK uses a heuristic procedure to determine which rules pointing to the expert's conclusion are closest to being satisfied in a misdiagnosed case. The procedures used by SEEK to automatically select experiments are somewhat complicated and we will only briefly touch on them in our discussion.

6.7 Overview of Model Refinement

Based on an empirical performance analysis of the rules, the SEEK system suggests refinements to a model's rules. However, even without this automatic advice, it should be easy to see how we may take advantage of the same form of empirical analysis to improve the performance of a model. Besides possibly changing the certainty weight of the conclusion in a rule, there are two qualitatively different ways in which we can modify a rule. The antecedent conditions of the rule can be *weakened* by removing components, thereby making the rule easier to satisfy. This is called a generalization of the rule. Alternatively, a rule's antecedent conditions can be *strengthened* by adding components. This is called a specialization and makes the rule more difficult to satisfy. The advice generated by the system consists of specific generalizations and specializations of the rules in a model. These are presented in the form of proposed experiments from which the model designer can select one to be subsequently tried, hoping that it will improve the model's performance. Some of the original work on pure learning, using the concept of generalization and specialization, can be found in Mitchell's "version space" learning scheme. Politakis has shown how these ideas can be used in a large-scale practical system with reasoning under uncertainty.

Changing the certainty weight in the conclusion of a rule can be also viewed as a generalization or a specialization. Increasing a conclusion's confidence produces a generalization because the rule will now have greater impact than before, when it gave a lower weight to the conclusion. In contrast, decreasing a rule's confidence reduces the impact of the rule since its conclusion will be weighted less. This, then, constitutes a specialization. Next we consider the task of generating rule refinement experiments. Figure 6.6 is an example of the process of analyzing potential rule refinements for hypothetical diagnosis Dx_1. In this example, the results of the model and the expert do not match for case 3. Because Dx_1 is the correct answer for case 3, performance might be improved by generalizing a Dx_1 rule, or alternatively, specializing a Dx_2 rule.

The analysis of a model's performance is central to the process of giving advice about rule refinement. Each case must be reviewed to determine if the

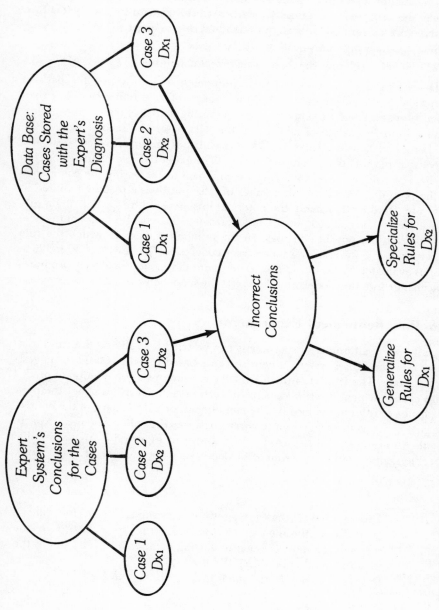

Figure 6.6:   Example of types of rule refinements for diagnosis Dx_1

model correctly diagnosed the case. A case is considered to be misdiagnosed by the model when the conclusion reached with greatest certainty by the model does not match the expert's conclusion stored with the case data. Two analyses are performed for each misdiagnosed case in order to:

1. generalize rules for the expert's conclusion
2. specialize rules for the model's incorrect conclusion.

If we tentatively assume that rule refinement is justified on the basis of empirical performance data alone, the process of rule refinement is a two-stage process. The first stage involves gathering statistics about rule performance from an analysis of all cases in the data base. The second stage is the evaluation of heuristic rules that use the statistics of performance to suggest specific experiments for rule refinement. The potential loss incurred in generalizing a rule by either removing a component or increasing the confidence is the possible increase in the number of false positives already attributed to the rule. On the other hand, the potential loss incurred by specializing a rule (by either adding a component or decreasing the conclusion's confidence) comes from the possible decrease in the number of cases in which the rule was responsible for reaching the correct conclusion. The merit of a modification is evaluated by performing an experiment that conditionally incorporates the change into the model, and then tests the model on the data base of cases.

6.8 Rule Refinement Experiments

Usually many alternative experiments can be tried for refining the rules in a model. In the SEEK system, a heuristic rule-based scheme is used to suggest experiments. The IF part of the heuristic rule contains a conjunction of predicate clauses which, if satisfied, confirm certain characteristics about the performance of rules in the model. The consequent, or THEN part, of the heuristic rule contains the suggestion to carry out a specific rule refinement experiment. An example of a heuristic rule is shown in Figure 6.7, suggesting the specific generalization experiment of decreasing the number of major findings.

If: The number of cases suggesting generalization of the rule is greater than the number of cases suggesting specialization of the rule and the most frequently missing component in the rule is the required number of "MAJOR" findings

Then: *Decrease the number of major findings in the rule.*

Figure 6.7:   Example of heuristic rule for suggesting an experiment

Evaluation of a heuristic rule begins by instantiating the clauses with the required empirical information about a specific rule in the model. If all clauses are satisfied, then the specific experiment is posted. The experiments suggested by the heuristic rules are narrowed by the expert to those changes that are consistent with his knowledge. In Figure 6.8, the experiments for the rules used in reaching the diagnosis of mixed connective tissue disease are presented after listing the misdiagnosed cases of this disease. The experiments are ordered based on the maximum potential performance gain that can be achieved on these cases. Other criteria for ordering could be considered, such as ease of change. For instance, an experiment that suggests changing the minors in a rule may be preferred over an experiment suggesting a change in majors. An explanation of a particular experiment is provided by translating the specific heuristic rule that was used to suggest the experiment into a narrative statement containing the empirical information about the rule. In Figure 6.9 we show the reasons that supported the first experiment. It should be emphasized that a decision as to which experiments, if any, are to be tried is left to the expert. Even though a particular experiment is supported empirically, an experiment must be justified in terms of knowledge about the domain. For example, is a rule resulting from the first experiment for rule 56 medically sound to make the diagnosis? The suggested experiments can lead the expert to reconsider the lists of major and minor findings for a particular final diagnosis and possibly to change or refine these findings. Because inexact heuristics are used, one is not absolutely certain of a net gain in performance before an experiment is tried.

24 cases in which the expert's conclusion MCTD does not match the model's conclusion:

1, 4, 11, 12, 14, 15, 42, 47, 49, 57, 60, 67,
71, 75, 78, 80, 84, 93, 99, 100, 104, 105, 107, 130

Proposed Experiments for Mixed Connective Tissue Disease

1. Decrease the number of majors in rule 56.
2. Delete the requirement component in rule 55.
3. Delete the requirement component in rule 54.
4. Decrease the number of minors in rule 57.
5. Delete the requirement component in rule 58.

Figure 6.8   Example of suggested experiments for rule refinement

:why(1)

If rule 56 had been satisfied, 8 currently misdiagnosed MCTD cases would have been diagnosed correctly. Currently, rule 56 is not used incorrectly in any of the cases. In rule 56 the component missing with the greatest frequency is the required number of "Major" findings.

Therefore, we suggest decreasing the number of majors in rule 56. This would generalize the rule so that it will be easier to satisfy.

Figure 6.9:   Example of support for an experiment

The results of a specific experiment are obtained by conditionally incorporating the revised rule(s) into the model. The updated model is then executed for all the cases in the data base. Suppose the model designer identifies an interesting experiment, such as the refinement of rule 56. Figure 6.10 illustrates a modification of the number of majors required by rule 56. The results for the change to rule 56 are summarized in Figure 6.11. In this example, the modification improves the performance significantly. Several misdiagnosed mixed connective tissue disease cases are now correctly diagnosed by the model. Moreover, there was no adverse side effect of this change on other cases with different final conclusions. The model designer has the option of either accepting or rejecting the experiment. If a simple modification does not yield desirable results, more complicated changes may be tried, such as multiple modifications or dropping a condition in a requirement.

Empirical analysis is important both in the design and validation of an expert system. Using a highly structured knowledge base, the SEEK system can use performance analysis to recommend experiments for rule refinement. We have given a brief overview of the operation of the SEEK system; some-

Rule 56 is:

3 or more Majors for MCTD
→ Possible mixed connective tissue disease

Generalization of Rule 56 is:

2 or more Majors for MCTD
→ Possible mixed connective tissue disease

Figure 6.10:   Example of rule modification

	Before		False Positives	After		False Positives
MCTD	9/ 33	(27%)	0	17/ 33	(52%)	0
Others	80/ 88	(91%)		80/ 88	(91%)	
Total	89/121	(74%)		97/121	(80%)	
Others						
RA	42/ 42	(100%)	9	42/ 42	(100%)	8
SLE	12/ 18	(67%)	4	12/ 18	(67%)	3
PSS	22/ 23	(96%)	5	22/ 23	(96%)	3
PM	4/ 5	(80%)	1	4/ 5	(80%)	1

Figure 6.11:   Before/after performance for a single rule refinement

what detailed heuristic procedures are used to allow the system to proceed beyond performance analysis to actually recommend changes in a model. Even without this automatic refinement capability, empirical performance information can be extremely valuable to the model designer. Ideally, an expert model abstracts the expert's reasoning, while the cases contain evidence about the accuracy of the model. An expert system developer must reconcile the results from these two sources of knowledge. The summarized performance results are a means for the expert to rethink the definition and structure of an expert model that is performing poorly for a specific sub-area. An analysis of the rules based on empirical case experience sharply focuses the expert's attention on modifications that can potentially improve performance. This can lead to a review of individual cases for inaccuracies in the data and to a reconsideration of the importance of specific criteria in the model. This process is not intended to custom-craft rules solely to the cases, but rather to provide the expert with explicit performance information that should prove helpful for modification of the rules.

In developing expert systems, we recognize that an expert's knowledge cannot be directly learned from sample cases except in extremely limited situations. In general, case review has been used to verify the performance of an expert model. If, however, we begin with both a set of expert-produced rules and an independent set of representative sample cases, we have the basis for a more valuable tool for rule refinement. With systems that integrate performance analysis into a model design framework, the expert will obtain a better formulation of the model and a better understanding of the explicit decision criteria that he uses in his reasoning.

6.9   Bibliographical and Historical Remarks

One of the first systems to advise a user on potential modifications to a
knowledge base to correct mistakes in an expert system was the TEIRESIAS
system reported in Davis (1979). The issue of performance analysis in an
expert system was addressed in Gaschnig (1979) with particular emphasis on
sensitivity analysis, i.e. how sensitive was the system to changes in evidence.
The notion of generalization and specialization (version spaces) in a learning
system for DENDRAL was explored in Mitchell (1982). The practical use of
empirical information to refine a knowledge base by suggesting rule modifications has been employed in the SEEK system (Politakis and Weiss 1984).

7

The Future of Expert Systems

7.1 Why Now?

The recent interest in the development of experts systems can be viewed as an interplay of both technical and social factors. Expert systems is a specialty within the broader field of artificial intelligence. Reviewing the recent past, we see that on a somewhat smaller scale, there was also a great deal of interest in developing intelligent systems in the 1960s. Many people proclaimed a new dawn for a computer age, where machines would soon exhibit highly intelligent characteristics. It was not unusual to hear predictions that the world chess champion would soon be a computer, and other more important tasks would soon be carried out or controlled by computer-based systems. Of course many of these predictions did not hold, and by the early 1970s artificial intelligence was considered to be a rather esoteric and isolated branch of computer science, scarred by the grandiose claims of the previous generation. Research was limited to a few universities, and research institutes which were funded for this relatively expensive variety of research by government agencies.

We must therefore try to place the present enthusiasm for expert systems into perspective. Artificial intelligence research is quite a young field, in its infancy compared to other scientific, engineering or even artistic endeavors. It is unrealistic to expect immediate practical results. Yet, as many researchers in the mid to late 1970s turned away from game-playing and "toy" problems and looked at real-world problems, they started noticing that they were achieving some degree of success in modeling expert behavior. While many projects began as basic research for knowledge representation, they evolved into realistic schemes for solving highly circumscribed problems. The R1 system is used routinely by Digital Equipment Corporation in processing orders for computer configurations. In one case, the PROSPECTOR system has been successful in identifying a mineral deposit which was previously overlooked. The Serum Protein Diagnostic Program instrument is used routinely in many hospital labs. Clearly, we see some definite technical achievement already, and the potential for developing even more powerful and imaginative expert systems is not hard to foresee.

Does this mean that developing expert systems is now routine? Hardly! The design and building of an expert system usually involves quite a difficult and intensive effort. The development of the expert system is almost always done under the close scrutiny of the expert, and brings together many skills and much knowledge from both computer science and the expert's specialty domain. A measure of performance must be obtained to be able to assert that a system is expert in some narrow sense. As the field becomes more popular, and given the inherent glamour of the word "expert," we are faced and will continue to be faced with misplaced claims for such systems. Many researchers have suddenly found the *right* terminology and are calling their systems expert, when in fact they are recycling already existing work under a new title. Because someone is building a system that will use domain knowledge, or builds a simple prototype, this hardly qualifies the resulting system as being expert. To some extent, then, we will be faced with some of the overextended claims we saw in the 1960s. The big difference is that we can expect more successes now than before, because we have a better idea of how to build systems that really do perform at an expert level, even though they do so only on well-circumscribed problems. Unfortunately, we do not know how to build these systems automatically, and therefore we need both skilled knowledge engineers and domain experts to work closely together in an effort which, like all human collaborations, can be quite fragile at times.

Besides the sheer technical knowledge gained about building expert systems, we have recently seen small and inexpensive computers, with markedly improved computational capabilities, come on the market. The cost of all types of computing has dropped dramatically and this trend is most likely to continue. Most developmental work in AI has taken place on relatively large time-shared computers. These machines are beyond the cost of many users, tend to become heavily loaded in the real-world, and therefore limit the applicability of most systems that may be developed. However, we are beginning to see small machines for relatively modest cost, with large address spaces (i.e. capable of handling large programs with no overlaying), some with virtual memories, more guaranteed (i.e non-competitive) CPU time, large amounts of disk space, and with high-speed graphics capabilities. Machines with some or all of these characteristics should not only enhance our ability to develop expert systems, but also increase the possibilities for their wider dissemination.

Yet, what may be the most important factor in increasing the acceptance of expert systems is society's changing view of computers. With the microcomputer revolution and the advent of personal computers, many people, particularly professionals, are changing their impressions of computers. They see them mostly as aids to increase their own productivity and creativity. The tasks performed by the computer may be simple time-saving efforts, such as personal financial planning, or they may be economically productive efforts, such as those in computer-aided design. Whereas a decade ago computers were viewed by the majority of people as a machine to be distrusted and possibly even feared, today they have become so familiar that TIME magazine changed the *Man of the Year* selection for 1982 to be a *Machine of the Year*. Computer programming has become mandatory for many students in their secondary education, and the number of jobs in computer-related industries makes these one of the high-growth fields. As more people perceive computers to be important and useful tools, they will come to view expert systems not as usurping their role, but rather as a form of "smart programming."

Not all people will welcome expert systems with open arms. To some, expert systems appear to be a direct threat. Many workers may feel that their jobs are threatened. This general argument is often cited as a negative factor in assessing the impact of computers on society. It is true that many people, particularly, those at the lower end of the economic scale, have been forced to retrain as new technologies displace old ones. The new jobs created by computerization typically require very different, more analytical skills than those needed by the smokestack industries now in decline. With expert systems, still other groups of individuals may feel threatened. Our society has become increasingly more specialized, with fast-paced developments being characteristic of many fields. Individuals may be protective of their experience and knowledge, unwilling to share it with others. Their knowledge, particularly if it involves tricks of the trade that have never been explicitly formulated, may be of great economic value to them. Perhaps the greatest benefit of expert systems is the potential of formalizing knowledge in areas where knowledge is mostly experiential and not widely disseminated, both because of economic reasons, and also because people may have no formal structure to give to decision-making knowledge. For example, one would think that for a field as highly developed as medicine, clear decision-making strategies for diagnosis would have evolved by now. Yet, although clinicians make many diagnoses daily, for many diseases, few formalized decision criteria are available. From a social point of view, expert systems are likely, in the next two decades, to help systematize the more well-established reasoning procedures used by experts. They will not replace the experts, but rather help people to move into more intellectually challenging activities where the knowledge encoded in an expert system is another routine source of information. On the positive side this ought to spur experts into more creative jobs, but given human nature, this is bound to meet resistance and resentment just as previous technological innovation has.

In this book we have examined almost exclusively the classification model approach. This is state-of-the-art technology in building expert systems. While other approaches may be used successfully, the classification model is well understood, and many systems empirically tested have used this approach. A number of generalized tools have been developed to help in the task of designing a prototype reasoning model. The typical classification model will use production rules and will cover a highly circumscribed problem. While the classification model may be a proven vehicle for building an expert system, the critical task of acquiring the knowledge from the expert and building the computer model remains in the hands of the skilled knowledge engineer. All too often people would like to think that powerful software and hardware are sufficient for building expert systems. Yet, for the foreseeable future what is needed is a cadre of talented people who know how to extract knowledge from an expert and represent that knowledge in a computer. Whether we call them knowledge engineers or scientists will depend on whether they confine their activities to system building or whether they perform more experimental work on the symbiosis of man/machine reasoning.

We recognize, of course, that not all problems can be represented by a classification model and alternative approaches may be necessary. These alternative approaches still present many open research issues because we have few comparable schemes with more complex structures that are both as generalizable and practical as the classification model. Our expectation is that many of the practical expert systems built in the near future will be based on the classification model. One may also reasonably expect that a particularly fruitful approach to designing more powerful yet practical expert systems is to start with the classification model and then consider extensions to the basic model which may be necessary to solve a particular problem. Obviously not all problems will be solvable. However, one of the hardest tasks in designing an application system is the initial specification of a reasoning model. The classification model will assist the designer in abstracting the expert knowledge, even when the representation does not ultimately turn out to be fully adequate. In many cases, the inadequacies may be removed by adding some new structures without removing the solid foundation of the classification model. There a number of areas in the development of expert systems where important advances can be made, but where perhaps more research will be needed before practical results can be expected. We will examine some of these areas next.

7.2   Future Research Directions

We have seen that the classification model using production rules can be an effective tool in describing expertise in a computer program. Several applications have been developed which demonstrate expert performance, at least in a narrow sense. Where then can we look to increase the performance and scope of expert systems, while decreasing the effort needed to build an expert system? While some may argue that the classification model is inherently simplistic, and therefore limited in potential, we would argue for extensions to the classification model, rather than a broad search for alternative representations. Here we are entering an area of informed opinion where, although some direction is available, the practical results to date do not support any one particular point of view over another. There are many research topics under investigation at various laboratories and institutions. We will describe those topics which seem to come up most frequently in our own experience with applications, and which therefore would have the greatest impact in advancing the class of problems for which solutions, as represented by expert systems, can be offered. Our view of these topics is slanted toward augmenting and improving the classification model.

7.2.1 ADDING NEW TYPES OF KNOWLEDGE TO THE SYSTEM

Clearly, we know that not all knowledge can be captured by simple associational rules represented in the computer as production rules. We also know that experts use other forms of knowledge in their reasoning. For example, many experts may use causal information, knowledge of structure or function, or mathematical relationships in solving their problems. None of these types of information is easily captured by production rules. Hart has addressed some of these issues by contrasting surface with deep models. The classification model would fall into the surface model category, whereas models capturing the other types of knowledge would be described as the deep model. Hart conjectures on the tradeoffs that he sees in developing expert systems based on these diverse sources of knowledge.

The surface model is relatively easy to represent and design. In contrast the deeper model will be much harder to describe and use in the early stages. As we try to increase the power and scope of the model, we see the limitations of the surface model and have difficulty in extending the performance of the model. Surely, we have an intuitive feeling that by having a richer set of relationships in the knowledge base, we will produce better systems. The problem with current state-of-the-art expert system models is that there are tradeoffs, since we acknowledge that the surface model is much easier to specify and use. We must be careful in recognizing that some problems may be simple enough for the surface model to be adequate, and even if we knew how to build the deep model, in the final analysis it may not be worth the effort. However, some problems may not be easily solvable as a classification problem, and will need to take advantage of the richer class of information that the expert may use. Even if it is not readily apparent that the expert uses such knowledge, we are faced with the tantalizing prospect of using this deep knowledge to exceed the performance of the expert. Today, however, we have no demonstrated generalized approaches or tools for the deeper models that are analogous to those which we have for the classification models. To us, the most promising approaches are those which start with the classification model and try to increase its capabilities by adding additional knowledge to the model in a controlled manner. For example, one might include mathematical relationships in a model by using already developed software and by communicating results between the surface model and this software. The hoped-for gain in such an approach is illustrated in Figure 7.1, which is derived from Hart's notions of deep versus surface systems. These ideas have been incorporated in the ELAS system which we described earlier.

7.2.2 EASING THE KNOWLEDGE ACQUISITION TASK

The ultimate design goal for knowledge acquisition is to to allow the expert to encode his own knowledge directly into the computer, removing the role of the knowledge engineer from the knowledge acquisition phase. Moreover, we would like the expert to spend as little time as possible in encoding his knowledge. We should not, however, expect to see any revolutionary change in the current balance between knowledge engineer and expert in the near future. Rather, we can expect a gradual evolution in the development of tools to facilitate the building of the knowledge base that will speed up the process

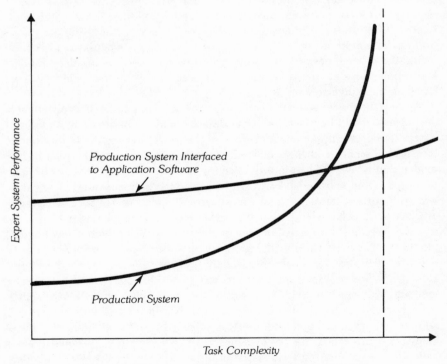

Figure 7.1: Increased performance of augmented production systems

of developing an expert system. We may see an increase in the number of people who can serve as knowledge engineers. It is not clear how easily this may be accomplished or how many people can qualify for such roles. As we have mentioned earlier, the skills required for the knowledge engineer are not the same as those required for programming, and these skills are usually developed during a period of apprenticeship to someone who is knowledgeable in the area. One of the best ways of simplifying the task of knowledge acquisition is to have a suitable framework for building prototypical models and a thorough understanding of the class of problems that are solvable or at least approachable using the given framework. This framework need not be complicated. We hope that the discussion in the previous chapters illustrated how one might build a relatively simple, but working system for encoding a set of production rules. If one is to address more than one problem, or one expects many changes to the knowledge base, it appears that both the knowledge acquisition task and the task of representing and encoding the knowledge can be simplified by having a general framework for capturing this knowledge. Although there is somewhat of a trend to specialized programming for each model, we still prefer the generalized tool approach, at least for designing a prototype. It is quite true that many problems may have specialized requirements, but often knowledge acquisition can proceed using a general framework with specialized modifications where necessary.

The knowledge acquisition phase of expert model development is particularly crucial to all other tasks. Improved hardware at lower costs will make expert systems more practical and more widespread. Improved software, particularly specialized knowledge engineering packages, can ease the task of knowledge representation. But for now, the role of the knowledge engineer is of prime importance. Despite the articles and advertisements that you will see extolling the virtues of various computers or AI languages, the knowledge engineer is an artist who can not be reproduced by supplying another individual with high quality paints and brushes.

7.2.3 LEARNING FROM EXPERIENCE

Most of the practical results in the near future are likely to come from applying current expert systems technology to real-world problems. This is also likely to lead to a better understanding of the types of problems we understand well versus those which need further research. This in turn will lead to gradual extensions to the basic classification model and an improved set of developmental tools. One area of great potential, but as yet unfulfilled promise, involves systems which learn from experience. Unfortunately, one has to learn to crawl before one walks, and even when we know what our experience has been, the knowledge engineer has a difficult enough time representing such information accurately in the computer. It is true that if one has a model that extensively uses certainty weights, it is possible to use experience in the form of cases to automatically update some of these numbers. However, as models grow large, containing hundreds of production rules, it is quite difficult to rely on such numbers for accurate results. The non-AI approaches, such as those used in statistics and pattern recognition, use parameters derived from gathered data, so that they can learn from experience by updating their statistics. In the expert systems approach, we model the expert and generally use experience as a means of verifying the data base. More significant learning for expert models would be, for example, to learn a new rule that could be added to the knowledge base. While an important area of research, improved results in "learning" will likely come about as we see more widespread and successful expert systems in operation. We are far from ready to supplant the knowledge engineer and expert with a system that automatically learns rules from experience and automatically builds the expert system itself.

7.2.4   INCORPORATING TIME RELATIONSHIPS INTO THE EXPERT MODEL

One important issue which has not been directly addressed yet is the influence of time relationships on decision-making. The classification model has no direct statement of time, although the observations used in the model could be made over time. Most classification models assume that we are looking at a snapshot in time and that all past information is summarized in the current snapshot. The situation can be quite complicated and there is much room for improvement in how time is incorporated in these models. While the snapshot summarizes the past, we know that this is not the usual way of gathering information. For example, in a medical setting, a patient may visit the physician over a period of days, weeks, months, or even years. On each visit observations will be made. Some of these observations will indicate changes in status and trends which must be taken into account by the reasoning model. In medicine, treatment management is highly dependent on these changes over time, such as the response to therapy. While we can use the idea of a snapshot which is constantly updated, the whole issue of how this extra dimension of time can be incorporated into a generalized classification model remains a research issue. Some results appear promising for certain specialized classes of models.

An issue related to time is "space." Most of the findings which have been discussed as part of the classification model are one-dimensional. However, in some cases a system must reason with multiple instances of the same item. The car repair model has been presented as a prototypical classification problem. A car may have many instances of the same object. For example, a car has four tires and many cylinders. If these multiple instances can be treated independently, there are direct solutions. Generalized solutions tend to grow more complex with increasing numbers of interactions between the multiple instances of objects. However, these problems may still be considered classification problems, as long as the set of hypotheses is relatively stable. Those problems, however, which require a system to generate hypotheses, i.e. the set of hypotheses cannot be listed a priori, tend not to be classification problems, and are usually more difficult to solve, such as problems which require searching among large numbers of possibilities as is the case in DENDRAL and many design problems. Most of the systems listed in Table 1.1 of Chapter 1 are classification-type systems.

7.3   The Future

The time appears to be at hand for expert systems to be widely used; most of the prerequisite accomplishments are in place. While this statement might have appeared radical only a few years ago, recent experience strongly supports this view. The process of diffusion will be gradual, and these programs may not end up being called expert systems, but perhaps viewed as smart programs for limited domains. Such programs will be based on domain knowledge of an expert and will perform with many of the attributes that we take to be expert when they are present in a human.

There is no question any more that hardware at reasonable costs will be available and will continue to improve. Marketplace trends show that many people, particularly professionals, increasingly view the computer as a problem-solving tool which can be exploited rather than as a threatening tool that will supplant them. Societies do not have the luxury of ignoring technological change because we live in a world economy where, if we do not change, the competition from other countries will force change on us. And it is always far better for us to design and control (to whatever extent possible) the changes than for others to do it for us! In limited applications it has already been demonstrated that programs can be designed with expert performance. These programs do not necessarily replace people, but serve as intelligent assistants, improving productivity and the quality of decision-making. We see, for example, that computer-aided design systems are readily accepted because these systems perform laborious tasks, such as assembling engineering drawings, which before took hours of human time, much of it given to repetitive mechanical tasks. Expert systems are a logical extension of the goals of such systems. While not limited to engineering problems, expert systems are an attempt to capture expert knowledge in a domain, allowing the computer to use this knowledge as an aid in problem solving. With all the necessary factors in place, we can forsee a bright future for such systems if the final ingredient, the knowledge engineer, becomes available in greater numbers. The knowledge engineer must still invest much time and effort in capturing the expertise in the computer. The cooperation of suitable experts will remain crucial. While we do not have the tools to automate these efforts, we can look forward to an gradual increase in the number of people who are qualified and interested in such collaborative efforts.

In the early years of computers, many people exaggerated the capabilities of these machines, claiming analytical capabilities that were far from real. Because of these unrealistic claims and expectations, a counter-reaction occurred and people became more skeptical of artificial intelligence. Because of the sheer number of changes in society and technology in past few years, many people are again looking for significant results from artificial intelligence research, particularly in the area of expert systems. This time it looks as if there is a solid foundation to these expectations.

7.4   Bibliographical and Historical Remarks

The relationship of expert systems to AI research and the relationship of artificial intelligence research to other sciences has recently been discussed (Nilsson 1982). Some promising directions for future research in AI and expert systems are described by Hart (1982). The successful determination of an overlooked mineral deposit by the PROSPECTOR system was reported in Campbell, Hollister, Duda, and Hart (1982). The development of a time-oriented expert system for an important medical problem was reported in Fagan (1978). Some practical generalized approaches for building expert systems that have capabilities for reasoning with time have been reported in Kastner, Weiss, and Kulikowski (1983).

References

(Aikins 1983) Aikins, J. Prototypical Knowledge for Expert Systems. *Artificial Intelligence* 20(2):163–210, February, 1983.

(Balzer, Erman, London, and Williams 1980) Balzer, R., Erman, L., London, P., and Williams, C. Hearsay-III: A Domain-Independent Framework for Expert Systems. In *Proceedings 1st Annual National Conference on Artificial Intelligence*, pages 108–110. Palo Alto, Ca., 1980.

(Barr and Feigenbaum 1981) Barr, A., and Feigenbaum, E. A. (editors). *Handbook of Artificial Intelligence*. William Kaufmann, Los Altos, Ca., 1981.

(Bennett and Engelmore 1979) Bennett, J., and Engelmore, R. SACON: A Knowledge-Based Consultant for Structural Analysis. In *Proceedings of the Sixth International Joint Conference on Artificial Intelligence*, pages 47–49. Tokyo, Japan, 1979.

(Bennett and Hollander 1981) Bennett, J., and Hollander, C. DART: An Expert System for Computer Fault Diagnosis. In *Proceedings of the Seventh International Joint Conference on Artificial Intelligence*, pages 843–845. Vancouver, Canada, 1981.

(Buchanan 1982) Buchanan, B. Research on Expert Systems. In Hayes, J., Michie, D., Pao, Y. (editors), *Machine Intelligence*, pages 269–300. Wiley, New York, 1982.

(Buchanan and Duda 1982) Buchanan, B., and Duda, R. *Principles of Rule-based Expert Systems*. Technical Report, Stanford University, Department of Computer Science, 1982.

(Buchanan and Feigenbaum 1978) Buchanan, B., and Feigenbaum, E. Dendral and MetaDendral: Their Applications Dimensions. *Artificial Intelligence* 11:5–24, 1978.

(Campbell, Hollister, Duda, and Hart 1982) Campbell, A., Hollister, V., Duda, R., and Hart, P. Recognition of a Hidden Mineral Deposit by an Artificial Intelligence Program. *Science* 217:927–929, 1982.

(Davis 1979) Davis, R. Interactive Transfer of Expertise: Acquisition of New Inference Rules. *Artificial Intelligence* 12:121–157, 1979.

(Davis, Austin, Carlbom, Frawley, Pruchnik, Sneiderman, and Gilreath 1981) Davis, R., Austin, H., Carlbom, I., Frawley, B., Pruchnik, P., Sneiderman, R., and Gilreath, J. The Dipmeter Advisor: Interpretation of Geologic Signals. In *Proceedings of the Seventh International Joint Conference on Artificial Intelligence*, pages 846–849. Vancouver, Canada, 1981.

(Duda and Hart 1973) Duda, R., and Hart, P. *Pattern Classification and Scene Analysis*. Wiley, New York, 1973.

(Duda and Shortliffe 1983) Duda, R., and Shortliffe, E. Expert Systems Research. *Science* 220:261–268, 1983.

(Duda, Gaschnig, and Hart 1979) Duda, R., Gaschnig, J., and Hart, P. Model Design in the Prospector Consultant System for Mineral Exploration. In Michie, D. (editor), *Expert Systems in the Microelectronic Age*, pages 153–167. Edinburgh University Press, Edinburgh, 1979.

(Duda, Hart, and Nilsson 1976) Duda, R., Hart, P., and Nilsson, N. Subjective Bayesian Methods for Rule-based Inference Systems. In *National Computer Conference*, pages 1075–1082. 1976.

(Duda, Hart, Barrett, Gaschnig, Konolige, Reboh, and Slocum 1978) Duda, R., Hart, P., Barrett, P., Gaschnig, J., Konolige, J., Reboh, R., and Slocum, J. *Development of the Prospector Consultation System for Mineral Exploration*. Technical Report, SRI, 1978.

(Fagan 1978) Fagan, L. *Ventilator Manager: A Program to Provide On-Line Consultative Advice in the Intensive Care Unit*. Technical Report HPP-78-16, Heuristic Programming Project, Stanford University, Stanford, Ca., September 1978.

(Feigenbaum 1977) Feigenbaum, E. The Art of Artificial Intelligence: I. Themes and Case Studies of Knowledge Engineering. In *Proceedings of the Fifth International Joint Conference on Artificial Intelligence*, pages 1014–1029. Cambridge, Mass., 1977.

(Forgy and McDermott 1977) Forgy, C., and McDermott, J. OPS, A Domain-Independent Production System Language. In *Proceedings of the Fifth International Joint Conference on Artificial Intelligence*, pages 933–939, 1977.

(Gaschnig 1979) Gaschnig, J. Preliminary Performance Analysis of the Prospector Consultant System for Mineral Exploration. In *Proceedings of the Sixth International Joint Conference on Artificial Intelligence*, pages 308–310. Tokyo, Japan, 1979.

(Genesereth 1982) Genesereth, M. Diagnosis Using Hierarchical Design Models. In *Proceedings of the Second Annual National Conference on Artificial Intelligence*, pages 278–283. Pittsburg, Pa., 1982.

(Hart 1982) Hart, P. Directions for AI in the Eighties. *SIGART Newsletter* (79):11–16, 1982.

(Hayes-Roth, Waterman, and Lenat 1983) Hayes-Roth, F., Waterman, D., and Lenat, D. (editors). *Building Expert Systems*. Addison-Wesley, New York, 1983.

(Kastner, Weiss, and Kulikowski 1983) Kastner, J., Weiss, S., and Kulikowski, C. An Efficient Scheme for Time-Dependent Consultation Systems. In *Proceedings MEDINFO 83: Fourth World Conference on Medical Informatics*, pages 619–622. North-Holland, Amsterdam, August, 1983.

(Kulikowski 1980) Kulikowski, C. Artificial Intelligence Methods and Systems for Medical Consultation. *IEEE Transactions on Pattern Analysis and Machine Intelligence* PAMI-2(5):464–476, September, 1980.

(Lindberg, Sharp, Kingsland, Weiss, Hayes, Ueno, and Hazelwood 1980) Lindberg, D., Sharp, G., Kingsland, L., Weiss, S., Hayes, S., Ueno, H., and Hazelwood, S. Computer-Based Rheumatology Consultant. In *Proceedings of the Third World Conference on Medical Informatics*, pages 1311–1315. North-Holland, 1980.

(MACSYMA Mathlab Group 1977) Mathlab Group. *MACSYMA Reference Manual*. Technical Report, Computer Science Laboratory, MIT, Cambridge, Mass., 1977.

(McDermott 1980) McDermott, J. *R1: A Rule-based Configurer of Computer Systems*. Technical Report CMU-CS-80-119, Carnegie-Mellon University, Department of Computer Science, Pittsburgh, Pa., 1980.

(McDermott 1982) McDermott, J. R1: A Rule-Based Configurer of Computer Systems. *Artificial Intelligence* 19:39–88, 1982.

(Miller, Pople, and Myers 1982) Miller, R., Pople, H., and Myers, J. INTERNIST-I, An Experimental Computer-based Diagnostic Consultant. *New England Journal of Medicine* 307:468–476, 1982.

(Mitchell 1982) Mitchell, T. Generalization as Search. *Artificial Intelligence* 18:203–226, 1982.

(Nau 1983) Nau, D. Expert Computer Systems. *Computer* 16:63–85, 1983.

(Newell and Simon 1972) Newell, A., and Simon, H. *Human problem solving*. Prentice-Hall, Englewood Cliffs, N. J., 1972.

(Nii and Aiello 1979) Nii, H., and Aiello, N. AGE (Attempt to Generalize): A Knowledge-based Program for Building Knowledge-Based Programs. In *Proceedings of the Sixth International Joint Conference on Artificial Intelligence*, pages 645–655. 1979.

(Nii, Feigenbaum, Anton, and Rockmore 1982) Nii H., Feigenbaum E., Anton J., and Rockmore A. Signal-to-Symbol Transformation: HASP/SIAP Case Study. *The AI Magazine* 3(2):23–35, Spring 1982.

(Nilsson 1980) Nilsson, N. *Principles of Artificial Intelligence.* Tioga, Palo Alto, Ca., 1980.

(Nilsson 1982) Nilsson, N., Artificial Intelligence: Engineering, Science, or Slogan. *The AI Magazine* 3(1):2–9, 1982.

(Pauker, Gorry, Kassirer, and Schwartz 1976) Pauker, S., Gorry, G., Kassirer, J., and Schwartz, W. Toward the Simulation of Clinical Cognition: Taking the Present Illness by Computer. *American Journal of Medicine* 60:981–995, 1976.

(Politakis and Weiss 1984) Politakis, P., and Weiss, S. Using Empirical Analysis to Refine Expert System Knowledge Bases. *Artificial Intelligence* in press, 1984.

(Reboh 1981) Reboh, R. *Knowledge Engineering Techniques and Tools in the Prospector Environment.* Technical Report Technical Note 243, SRI International, 1981.

(Shortliffe 1976) Shortliffe, E. *Computer-Based Medical Consultations: MYCIN.* Elsevier Scientific Publishing Company, Inc., New York, 1976.

(Shortliffe and Buchanan 1975) Shortliffe, E., and Buchanan, B. A Model of Inexact Reasoning in Medicine. *Mathematical Biosciences* 23:351–375, 1975.

(Shortliffe, Buchanan, and Feigenbaum 1979) Shortliffe, E., Buchanan, B., and Feigenbaum, E. Knowledge Engineering for Medical Decision-Making: A Review of Computer-based Clinical Decision Aids. *Proceedings of the IEEE* 67:1207-1224, 1979.

(Szolovits 1982) Szolovits, P. (editor). *Artificial Intelligence in Medicine.* Westview Press, Boulder, Colo., 1982.

(Szolovits and Pauker 1978) Szolovits, P., and Pauker, S. Categorical and Probabilistic Reasoning in Medical Diagnosis. *Artificial Intelligence* 11:115–144, 1978.

(Van Melle 1979) Van Melle, W. A Domain-independent Production-Rule System for Consultation Programs. In *Proceedings of the Sixth International Joint Conference on Artificial Intelligence,* pages 923–925. Tokyo, Japan, 1979.

(Weiss and Kulikowski 1979) Weiss, S., and Kulikowski, C. EXPERT: A System for Developing Consultation Models. In *Proceedings of the Sixth International Joint Conference on Artificial Intelligence,* pages 942–947. Tokyo, Japan, 1979.

(Weiss, Kulikowski, Amarel, and Safir 1978) Weiss, S., Kulikowski, C., Amarel, S., and Safir, A. A Model-based Method for Computer-aided Medical Decision-Making. *Artificial Intelligence* 11(1,2):145–172, August, 1978.

(Weiss, Kulikowski, and Galen 1981) Weiss, S., Kulikowski, C., and Galen, R. Developing Microprocessor-based Expert Models for Instrument Interpretation. In *Proceedings of the Seventh International Joint Conference on Artificial Intelligence,* pages 853–855. Vancouver, Canada, 1981.

(Weiss, Kulikowski, and Safir 1978) Weiss, S., Kulikowski, C., and Safir, A. Glaucoma Consultation by Computer. *Computers in Biology and Medicine* (1):25–40, 1978.

(Weiss, Kulikowski, Apte, Uschold, Patchett, Brigham, and Spitzer 1982) Weiss, S., Kulikowski, C., Apte, C., Uschold, M., Patchett, J., Brigham, R., and Spitzer, B. Building Expert Systems for Controlling Complex Programs. In *Proceedings of the Second Annual National Conference on Artificial Intelligence,* pages 322–326. Pittsburg, Pa., 1982.

(Yu et al. 1979) Yu, V., Fagan, L., Wraith, S., Clancey, W., Scott, A., Hannigan, J., Blum, R., Buchanan, B., Cohen, S., Davis, R., Aikins, J., Van Melle, W., Shortliffe, E., Axline, S. Antimicrobial Selection for Meningitis by a Computerized Consultant: A Blinded Evaluation by Infectious Disease Experts. *Journal of the American Medical Association* 241:1279–1282, 1979.

(Zadeh and Fukanaka 1975) Zadeh, L., and Fukanaka, K. (editors). *Fuzzy Sets and their Applications to Cognitive Decision Processes.* Academic Press, New York, 1975.

Author Index

Subject Index